Wastewater Collection System Operator Certification Studybook

Second Edition

By

James J. Courchaine

2019

Water Environment Federation
601 Wythe Street
Alexandria, VA 22314-1994
www.wef.org

IMPORTANT NOTICE

About WEF

The Water Environment Federation (WEF) is a not-for-profit technical and educational organization of 34,000 individual members and 75 affiliated Member Associations representing water quality professionals around the world. Since 1928, WEF and its members have protected public health and the environment. As a global water sector leader, our mission is to connect water professionals, enrich the expertise of water professionals, increase the awareness of the impact and value of water, and provide a platform for water sector innovation. To learn more, visit **www.wef.org**.

CONTENTS

PREFACE

In many parts of North America and other areas around the world, utilities, government agencies, wastewater associations, and elected officials are increasingly recognizing the importance and need for properly trained and certified wastewater collection system operators. This recognition is reflected in the many voluntary and mandatory certification requirements for wastewater collection system professionals. These certification programs require that operators pass written examinations that typically cover a wide range of management and operations and maintenance topics.

The purpose of this study book is to help wastewater collection system operators expand knowledge in their chosen field and assist them in preparing for certification examinations. Experience has shown that practice with sample examination questions is extremely beneficial in preparing for these tests.

The questions presented here cover the following broad categories: background knowledge, support systems, operation and maintenance, supervision and management, safety procedures, design and construction, electrical pumps and motors, and mathematics. Developing strong mathematical skills will significantly help an operator pass certification examinations and make them more efficient and effective operators. To instruct in this area, detailed solution sets are provided for mathematical problems. All numerical problems are presented in both U.S. customary and International System (SI/metric) units. A comprehensive glossary and reference list are also provided.

The questions presented here were compiled, reviewed, field tested, and edited over the last four decades by the author, James J. Courchaine. They are based on information included in standard reference texts, including those published by the Water Environment Federation (WEF), and on sample questions supplied by certification agencies from the United States and Canada.

The questions were written and reviewed by collection system professionals who are members of WEF and/or of the New England Water Environment Association (NEWEA).

This publication is an outgrowth of an earlier collection of questions that was developed jointly with Donald S. Pottle. Like the author, Don Pottle is a past chair of NEWEA's Collection System Operator Certification Committee, and his role in laying the groundwork for the development of this study book is gratefully acknowledged.

ABOUT THE AUTHOR

James J. Courchaine has more than 45 years of experience in the management and operations and maintenance of water distribution, water and wastewater treatment, and wastewater collection systems. He has been directly involved in the training and certification of collection system operators for more than 40 years. Mr. Courchaine developed basic and advanced courses in the management, operations, and maintenance of wastewater collection systems for the University of Lowell in Massachusetts and taught these courses for more than 10 years. He is certified as an operator in New England for wastewater treatment and collection systems operation and in Massachusetts for water treatment and distribution. Additionally, he has held high-level management positions with responsibility for the water, wastewater, stormwater, and public works operations of several communities in New England and Arizona.

Mr. Courchaine has been active in the New England Water Environment Association (NEWEA), particularly in the field of certification and training, for almost 40 years. He has served as a NEWEA officer in several key positions for several terms, including president in 2004. He has received numerous awards in leadership, management, operations, and maintenance, which include the NEWEA James J. Courchaine Collection Systems Award for outstanding efforts in the management, operation, maintenance, design, and safety of wastewater collection systems; the NEWEA Alfred E. Peloquin Award for exemplary performance and significant contributions to the wastewater field; the Collection Systems Golden Hook Award from his peers in recognition of contributions in personnel development and training; the Water Pollution Control Federation (WPCF) International Collection Systems Award for practical application of an original concept that expands knowledge of wastewater collection; the WPCF E. Sherman Chase Award for outstanding meritorious service in the wastewater collection systems field; the Founders Award in recognition of long-term contributions to NEWEA and dedicated service to the industry in any facet of water pollution control, notable achievements, or accomplishments in the environmental field; the NEWEA and the New Jersey Water Environment Association (NJWEA) 5 S Award; the NEWEA, NJWEA, and WEF Golden Manhole Society; and the California Water Environment Association (CWEA) Pick Award.

While working with the Massachusetts Department of Environmental Protection, Mr. Courchaine managed the state inflow/infiltration (I/I) program. He has also held positions with several nationally known consulting firms and is currently Vice President and National Director of Business Practices for Tata & Howard. He has served on the Collection Systems Committees of NEWEA, CWEA, and WEF for many years. Jim has worked with NEWEA Wastewater Collection System Certification Committee in the development of NEWEA's very successful voluntary certification program that has certified more than 6500 operators since its inception in 1980. He has also volunteered his time and provided many questions for the wastewater collection system exams of the Association of Boards of Certification.

ABOUT THE REVIEWERS

Rick Arbour has more than 50 years of operation and maintenance experience, including 10 years in U.S. industry; 15 years in a large, regional public utility; and more than 25 years as a consultant. He is a nationally recognized expert in all aspects of operation, maintenance, and management of water and wastewater utilities. His clients have included large consulting firms, law firms, water and wastewater utilities, federal and state regulatory agencies, universities, international corporations, and international funding agencies. He has provided services on projects throughout the United States, Canada, Mexico, and South America.

Mr. Arbour has developed and delivered certification training courses for water distribution and wastewater collection systems for the Minnesota Pollution Control Agency, Regional Municipalities of Ottawa and Niagara, and the Mexican Institute of Water Technology. He has authored or revised publications on collection system operation, maintenance, and management for California State University, Sacramento; the Water Environment Federation (WEF); and the Minnesota Pollution Control Agency. He has served as a continuing education faculty member in the field of sanitary sewer operations, maintenance, and management at the University of Wisconsin, Madison, for 18 years. Mr. Arbour has developed and delivered 2- to 3-day collection system seminars on the operation, maintenance and management of pumping stations and gravity sewer systems at University of Nevada, Las Vegas, for 13 years. He has developed and delivered annually a 2-day wastewater collection system Class SA and SB certification training seminar at the Minnesota Pollution Control Agency sponsored Annual Collection System Operators Conference. He is a Past Chair of WEF's Collection Systems Committee and is a WEF Life Member. He has held the highest level of certification in Minnesota since 1975 and is certified as a Grade IV operator in New England.

William Di Tullio, Jr., P.E., has a Bachelor of Science Degree in Civil Engineering and a master's degree in Environmental Engineering from Northeastern University. He is a Certified Grade IV Collection System Operator. Mr. Di Tullio is a member of the Northeast Trenchless Technology Society Board of Directors and an active member of WEF. Mr. Di Tullio serves as Malcolm Pirnie's Technical Director for Performance Enhancement related to Buried Infrastructure Management and Infrastructure System Evaluations. He served as chairman of the 1991 Massachusetts Department of Environmental Protection (MADEP), Infiltration/Inflow Guidelines Update Committee and as a co-author for WEF Manual of Practice No. FD-17 titled *Combined Sewer Overflow Pollution Abatement* (WEF, 1999), and MADEP Manual of Guidance for Operation and Maintenance of Collection Systems.

Rudy Fernandez, P.E., has been a civil engineer since 1972, working primarily on infiltration and inflow (I/I) projects for wastewater collection systems. He is a principal at RJN Group, where he has worked since 1982. Mr. Fernandez has managed engineering projects throughout the East Coast of the United States, performing I/I studies, rehabilitation design, and construction management. He is active in WEF's Collections Committee, having served as its chair and as chair of the Collection Systems Subcommittee of WEF's Program Committee and Technical Practice Committee. He has also served as chair of the Florida Water Environment Association's Collection System Committee. Mr. Fernandez co-authored an appendix for the 1984 edition of WEF's *Confined Space Entry*. Mr. Fernandez is a licensed professional engineer in several East Coast states.

George Martin has been involved in the management, operations, and maintenance of wastewater collection systems since 1982. He was a founder of the South Carolina Voluntary Certification Committee that has certified approximately 1700 wastewater collection system operators to date. Mr. Martin holds certificate number 0001 for the collections systems operator certification program in the State of South Carolina. He was instrumental in establishing a 3-day school held twice each year that provides training for wastewater collection system operators. Mr. Martin is a past president of the Water Environment Association of South Carolina, has chaired several technical committees in the state, and has served as chair of WEF's Collection Systems Committee. He has spoken on many collection systems topics, including best management practices for management, operations, and maintenance throughout the country.

1 BACKGROUND KNOWLEDGE

Most people who choose to work in the wastewater collection system profession initially have no formal education in the profession. Rather, they gain their knowledge through "on-the-job" training. Proper management, operation and maintenance, and safety of a wastewater collection system are essential to protect the public health, the wastewater collection system asset, and its operators. To understand how wastewater collection systems should function efficiently and effectively, a basic knowledge of infrastructure management, operations and maintenance, and hydraulics is necessary. This chapter covers fundamental questions about wastewater collection system components management, operations and maintenance, and safety procedures.

1. A flow of 60 gph (227 L/h) is equal to

    a) 1 mgd (3.785×10^3 m³/d)
    b) 1 mg/L (1 mg/L)
    c) 10 gpm (0.63 L/s)
    d) 1 gpm (0.063 L/s)

2. The slope of a sewer can be measured by

    a) The difference from the crown to the invert
    b) The difference in invert elevations over a length of pipe
    c) Using the diameter of the pipe
    d) Only entering the manhole

3. An air-gap can be used to

    a) Clean sewer blockages caused by grease
    b) Let air into manholes
    c) Prevent backflow of wastewater into a drinking water supply
    d) Test for oxygen deficiency in manholes

4. A cross-connection is a

   a) Cast iron pipe connected to a concrete pipe
   b) Steel pipe connected to an asbestos pipe
   c) Potable water pipe connected to a supply of questionable origin
   d) Lateral line connected to an interceptor

5. Wastewater carried in a collection system may come from

   a) Digester supernatant
   b) Exfiltration
   c) Primary settling tank
   d) Inflow

6. When using any hydraulic sewer cleaning method, care must be taken by the operator

   a) To always plug the downstream manhole
   b) Not to cause flooding in homes and businesses
   c) To prevent any air gap from occurring
   d) To throttle flows from the hydrants by using the hydrant valve only

7. Disease-producing bacteria are described as

   a) Saprophytic
   b) Facultative
   c) Pathogenic
   d) Parasitic

8. A manhole with a center grating is typically used for

   a) Venting
   b) Storm sewers
   c) Sanitary sewers
   d) None of the above

9. A pig would be used as _____ in a force main

   a) A reamer
   b) A stopper
   c) A float
   d) A tell-tale

10. A minimum scouring velocity of 2 ft/sec (0.6 m/s) in sewers is necessary so that

    a) Flow velocities can be estimated between manholes

    b) Roots will not grow in sewers

    c) Sewer pipes will not become eroded on the bottom

    d) Solids will not build up in a sewer and reduce flow capacity

11. A cross-connection is best defined as

    a) Groundwater entering a gravity sewer through cracks in the piping

    b) A lateral line connected to the main illegally

    c) Storm drainage piped into a domestic collection system

    d) A potable water supply connected to a potential source of pollution

12. When dealing with collection systems, *I & I* refers to

    a) Inflow and inspection

    b) Inspection and information

    c) Infiltration and inflow

    d) None of the above

13. Corrosion of sewer pipes may be caused by

    a) Laminar flow through them

    b) Oils in wastewater

    c) Fungi

    d) Acids in wastewater

14. In a collection system, wastewater that contains no "free" or dissolved oxygen is referred to as

    a) Aerobic

    b) Anaerobic

    c) Ambient

    d) Debris

15. The primary reason why it is important *not* to have cross-connections of sewers with water supply systems is

    a) Possible overloading of water systems

    b) Possible loss of drinking water pressure

    c) Possible drinking water contamination

    d) Loss of wastewater from the sewerage system

16. A pig would most likely be used in the cleaning of a

    a) Gravity sewer
    b) Storm sewer
    c) Force main
    d) House or building sewer

17. Flowrates can be expressed in which of the following units of measurement?

    a) in.^{3} (mm^3)
    b) gal/lb (mL/kg)
    c) mg (L/s)
    d) lb/mil. gal (mg/L)

18. A sewer manhole is

    a) The lowest point in the channel
    b) A gate that opens or closes swinging around a set of hinges
    c) An opening in a sewer provided for the purpose of access
    d) A small hole in a sewer where a wastewater service line connects

19. The crown of a sewer pipe is the

    a) Bottom
    b) Top
    c) Flow line
    d) Outside at the grade line

20. The primary purpose of wastewater disinfection is

    a) Reduction of suspended solids
    b) Destruction of algae
    c) Prevention of receiving water contamination
    d) Oxidation of effluent

21. A flow of 1 000 000 gpd (3785 m^3/d) is approximately

    a) 500 gpm (approximately 32 L/s)
    b) 700 gpm (approximately 44 L/s)
    c) 1000 gpm (approximately 63 L/s)
    d) 60 000 gph (approximately 63 000 mL/s)

22. What should be considered as the scouring velocity of sanitary sewers?

    a)   1 ft/sec (0.3 m/s)
    b)   2 ft/sec (0.6 m/s)
    c)   4 ft/sec (1.2 m/s)
    d)   4.5 ft/sec (1.4 m/s)

23. A manhole barrel is referred to as

    a)   An object with flashing lights placed next to a manhole for safety
    b)   The entrance to the manhole
    c)   The cylindrical section between the cone and the shelf
    d)   The cover support around the rim

24. Of the following, which does *not* lead to roots entering a collection system?

    a)   Improper pipe bedding
    b)   Misaligned pipe joints
    c)   Differential settling
    d)   Manhole walls properly sealed

25. Of the following abbreviations listed, which is *not* used in computing or expressing flowrates?

    a)   ft/sec (m/s)
    b)   gpm (L/s)
    c)   mgd (m^3/d)
    d)   psi (kgf/cm^2)

26. Sewer lines can move after years of being in place because of

    a)   Settlement
    b)   Frost action
    c)   Nearby activities
    d)   All of the above

27. The end of a sewer pipe that is formed to fit into the bell of the next pipe is called a/an

    a)   Clamp
    b)   Spigot
    c)   Invert
    d)   Housing

28. Which of the following is a component of a manhole?

    a) Cone

    b) Barrel

    c) Shelf

    d) All of the above

29. When the velocity of flow in a collection system is 2 ft/sec (0.6 m/s) or greater,

    a) A hydraulic jump will occur

    b) A scouring action will develop

    c) Dissolved solids will separate causing a drop in pH

    d) Solids will settle out and cause blockage in the sewer line

30. Inflow can best be defined as

    a) Flow into a treatment facility

    b) Direct discharge of stormwater to a sewer

    c) Storm flow plus wastewater flow

    d) None of the above

31. Solids that may settle to the bottom of a wet well are called

    a) Colloidal solids

    b) Scum

    c) Suspended solids

    d) Settleable solids

32. Odors encountered in manholes at the end of force mains are usually caused by

    a) Scavengers

    b) Chemicals in the wastewater

    c) Aerobic bacteria

    d) Anaerobic conditions

33. _____ is a disease that can be caused by the ingestion of wastewater.

    a) Influenza

    b) Tetanus

    c) Typhoid

    d) Legionnaire's disease

34. The gas most typically associated with septic wastewater is

    a)   Carbon monoxide

    b)   Carbon dioxide

    c)   Hydrogen sulfide

    d)   Methane

35. The recommended maximum allowable velocity in sewers is

    a)   1 ft/sec (0.3 m/s)

    b)   5 ft/sec (1.5 m/s)

    c)   8 ft/sec (2.4 m/s)

    d)   10 ft/sec (3.1 m/s)

36. In addition to being poisonous, hydrogen sulfide can also cause which of the following to happen in a collection system?

    a)   Vapor locks in manholes

    b)   Stoppages in sewer lines

    c)   Crown corrosion

    d)   All of the above

37. *Infiltration* can best be defined as

    a)   The flow of groundwater into a sewer through faulty joints

    b)   Industrial waste discharge into a sanitary sewer

    c)   Stormwater discharged into a sanitary sewer

    d)   All of the above

38. A lift station operations and maintenance manual should include

    a)   Names and emergency telephone numbers of collection system operators, vendors, and contractors

    b)   An agency budget

    c)   Instructions for preserving biochemical oxygen demand (BOD) samples

    d)   None of the above

39. The purpose of an air-gap device is to

   a)   Eliminate the need for check valves in lift stations

   b)   Prevent any water of questionable quality from contaminating potable water sources

   c)   Allow storm flows to bypass the collection system and allow stormwater to flow over outfalls

   d)   Reduce corrosion of manhole covers and rungs

40. In a collection system that serves mainly residential customers, at what time of day would lowest flows generally occur?

   a)   Midmorning

   b)   Afternoon

   c)   Evening

   d)   Late night

41. A combined sewer would be best defined as

   a)   A sewer intended to receive both wastewater and storm and/or surface water

   b)   A sewer combined with other sewers

   c)   A sewer line solely used to collect stormwater

   d)   A sewer line solely used to collect infiltration and inflow

42. An invert is best defined as

   a)   Water level

   b)   Sea level

   c)   Lowest point in the channel

   d)   Highest point in the channel

43. Infiltration can be caused by

   a)   Poor joints

   b)   Cracked pipes

   c)   Root intrusion

   d)   All of the above

44. Which of the following wastewater equipment is *not* considered a part of the collection system?

    a) Cleanouts
    b) Bar screens
    c) Lift stations
    d) None of the above

45. The abbreviation "mgd" means

    a) Metric gallons per day
    b) Maximum gallons per day
    c) Mean gallons per day
    d) Million gallons per day

46. The pH of a solution is based on

    a) The temperature
    b) The dissolved material
    c) The hydrogen ion concentration
    d) The concentration of a Na^+ solution

47. When there is a collection system failure, what is likely to happen?

    a) Wastewater gets treated
    b) The workers fail to get paid
    c) Public safety may be compromised
    d) The public will not notice anything

48. During a rainstorm, which of the following may increase the most?

    a) Exfiltration
    b) Inflow
    c) Exflow
    d) They will all increase at the same rate

49. Influent can best be defined as

    a) The flow of wastewater into a stream
    b) The discharge of industrial waste into a sewer
    c) The flow of wastewater entering the collection system
    d) a and b

50. One million gallons per day (3785 m³/d) is approximately equal to

a)   1.0 cu ft/sec (28.3 L/s)
b)   1.5 cu ft/sec (43.8 L/s)
c)   10 cu ft/sec (283.1 L/s)
d)   15 cu ft/sec (424.7 L/s)

51. *BOD* is defined as

a)   Backhoe operating depth
b)   Bacteria optical density
c)   Biochemical oxygen demand
d)   None of the above

52. Heavier material and settleable solids are removed from wastewater in

a)   Vacuum filtration
b)   Primary treatment
c)   Preliminary treatment
d)   None of the above

53. Which of the following *best* describes how the largest quantity of water infiltrates sanitary sewers?

a)   Leaks through manhole covers
b)   Porous pipes
c)   Defective pipe joints
d)   Drainage through catch basins

54. Of the following choices, only one is *not* used to convey wastewater from homes or businesses. That one exception is

a)   Building sewer
b)   Lateral sewer
c)   Storm sewer
d)   None of the above

55. High concentrations of heavy metals should be precluded from entering the sanitary sewer system by industrial pretreatment requirements because

   a)   They will damage pumps and motors

   b)   They tend to cause sewer stoppages

   c)   They will eventually improve the receiving waters

   d)   They may affect the treatment process

56. A medium strength wastewater contains approximately

   a)   50 to 75 mg/L BOD_5 and suspended solids

   b)   75 to 100 mg/L BOD_5 and suspended solids

   c)   150 to 300 mg/L BOD_5 and suspended solids

   d)   400 to 800 mg/L BOD_5 and suspended solids

57. Which is *best* used to determine when a sewer line needs to be cleaned?

   a)   The amount of free time the crew has

   b)   Odors coming from manholes

   c)   Blockages and complaints

   d)   Experience and records

58. One cubic yard (cu yd) (0.765 m³) of concrete contains

   a)   3 cu ft (85 L)

   b)   9 cu ft (255 L)

   c)   27 cu ft (765 L)

   d)   33 cu ft (934 L)

59. The principal purpose of a standard manhole is to

   a)   Monitor flow

   b)   Trap rodents

   c)   Connect roof drains

   d)   Provide access for cleaning and inspection

60. Storm drains are intended to carry

   a)   Sanitary wastewater only

   b)   Surface and storm water

   c)   Both stormwater and wastewater

   d)   All of the above

61. *Peak flow* refers to

a)   Highest BOD level
b)   The most corrosive period of flow
c)   Highest flow during a certain period of time
d)   The design flow of the sewer

62. You would expect to find a *drop* manhole at the

a)   Summit of a sewer
b)   Intersection of two streets
c)   Junction of two sewers of different levels
d)   Junction of two sewers of different diameters

63. The invert of a sewer pipe is the

a)   Top
b)   Outside diameter
c)   Outside at the grade line
d)   Bottom where wastewater flows

64. The elevation of a sewer refers to the elevation of the

a)   Center line of the sewer
b)   Center of the top on the inside
c)   Center of the top on the outside
d)   Center of the bottom on the inside

65. If the inside diameter of a sewer pipe is 20 in. (500 mm) and the pipe wall is 1.5-in. (40-mm) thick, what is the outside diameter of the pipe?

a)   21.5 in. (540 mm)
b)   22.5 in. (570 mm)
c)   23.0 in. (580 mm)
d)   24.0 in. (610 mm)

66. Odors from wastewater collection systems *typically* result from

a)   High alkaline wastewater
b)   Fresh domestic wastewater
c)   Infiltration of alkaline ground water
d)   Anaerobic conditions developing in solids accumulated in the pipe

67. Why is exfiltration a possible danger to public health?

    a)   Pollution of well water is possible

    b)   Significant increase in groundwater levels could occur

    c)   It can lead to unacceptably high flows in sewer systems

    d)   None of the above

68. Which two of the following factors most generally determine the capacity of a sewer?

    (1)   Depth

    (2)   Size

    (3)   Slope

    (4)   Pipe material

    a)   1 and 2 above

    b)   1 and 3 above

    c)   2 and 3 above

    d)   2 and 4 above

69. Wastewater is treated to

    a)   Prevent pollution

    b)   Protect public health

    c)   Remove harmful wastes from wastewater

    d)   All of the above

70. One milligram per liter (1 mg/L) can also be expressed as

    a)   0.01%

    b)   0.001%

    c)   0.0001%

    d)   0.00001%

71. The average person contributes _____ of suspended solids to the sewer daily.

    a)   0.14 lb (0.06 kg)

    b)   0.20 lb (0.09 kg)

    c)   0.38 lb (0.17 kg)

    d)   0.50 lb (0.23 kg)

72. Three types of sewers are

    a)   Sanitary, storm, pipes

    b)   Sanitary, storm, combined

    c)   Sanitary, storm, conventional

    d)   Conventional, surface, combined

73. A stick travels 60 ft (18.3 m) in 40 sec in a 10-in. (250-mm) sewer. What is the flow velocity in the sewer?

    a)   0.5 ft/sec (0.15 m/s)

    b)   0.67 ft/sec (0.20 m/s)

    c)   1.0 ft/sec (0.30 m/s)

    d)   1.5 ft/sec (0.46 m/s)

74. Which of the following is *not* used in measuring flow?

    a)   Weir

    b)   Ball and weight

    c)   Venturi meter

    d)   Parshall flume

75. Given the temperature of wastewater at 50 °F, what is the corresponding temperature in °C?

    a)   10 °C

    b)   12 °C

    c)   17 °C

    d)   20 °C

76. Excessive flows encountered during a storm are an indication of

    a)   High ratio of exfiltration

    b)   High rates of inflow

    c)   Abnormally high water usage

    d)   All of the above

77. The angle of repose is defined as

    a)   The top of the soil minus the distance to the trench

    b)   Minimum natural slope

    c)   Maximum natural slope

    d)   The depth from the top of the soil to the bottom of the trench

78. A trunk sewer is

    a)   A main sewer line that receives many laterals

    b)   A main sewer line that receives many branches

    c)   A main sewer line that serves a large territory

    d)   All of the above

79. Exfiltration is a problem in wastewater collection systems because

    a)   It can cause excessive flow problems at the water resource recovery facility

    b)   It may contaminate an aquifer

    c)   Valuable bugs could be lost that would aid in the treatment process

    d)   None of the above

80. Actual velocities in sewers may be measured by

    a)   Time of travel using dye

    b)   Time of travel using a float

    c)   Time of travel using a radioactive tracer

    d)   All of the above

81. A drop manhole is

    a)   A manhole where the outgoing pipe invert is higher than the influent line

    b)   A manhole where the outgoing pipe invert is significantly lower than the influent pipe invert

    c)   A manhole that has more than one influent pipe

    d)   A manhole with a sump

82. The gauge pressure that is equivalent to the static head from a 75-ft (22.9-m) column of water is approximately

    a)   9.1 psig (62.7 kPa)

    b)   20.0 psig (137.9 kPa)

    c)   32.5 psig (224.1 kPa)

    d)   48.8 psig (336.5 kPa)

83. Wastewater is typically

    a)   10% solids

    b)   25% settleable solids

    c)   12% volatile

    d)   99.9% water

84. When crossing under a stream, which type of pipe is most commonly used?

    a)   Ductile iron
    b)   Concrete
    c)   Class 52 fiberglass
    d)   Blue brute class 52

85. One cubic foot (cu ft) (0.028 m³) of water weighs approximately

    a)   8.34 lb (3.8 kg)
    b)   14.7 lb (6.7 kg)
    c)   62.4 lb (28 kg)
    d)   None of the above

86. Wastewater temperature is reported to be 12 °C. What is the temperature in °F?

    a)   48.9 °F
    b)   53.6 °F
    c)   58.1 °F
    d)   58.4 °F

87. The minimum velocity to prevent solids deposition in sewers is approximately

    a)   1 ft/sec (0.3 m/s)
    b)   2 ft/sec (0.6 m/s)
    c)   5 ft/sec (1.5 m/s)
    d)   10 ft/sec (3.1 m/s)

88. The ingestion of wastewater can cause which of the following diseases?

    a)   Sinusitis
    b)   Tetanus
    c)   Bronchial pneumonia
    d)   None of the above

89. An equation that describes the gravity flow of wastewater in sewer pipes is

    a)   Chick's Law
    b)   The Chezy equation
    c)   The Manning equation
    d)   The Hazen-Williams equation

90. The hydraulic radius of a sewer refers to

    a)   The diameter

    b)   The perimeter

    c)   One-half the diameter

    d)   The wetted perimeter divided by the diameter

91. If you are informed that a large quantity of acid waste was accidentally discharged into a sanitary sewer in your city, you could determine when it reached the water resource recovery facility because

    a)   The pH of the raw wastewater would rise

    b)   The pH of the raw wastewater would fall

    c)   The volatile portion of the suspended solids would be greatly increased

    d)   The dissolved oxygen in the raw wastewater would exceed the saturation value

92. *Drawdown* is a hydraulic term that refers to the following condition in a typical sewer:

    a)   Siphoning

    b)   Draining the line

    c)   An area where the flow slows

    d)   The surface curve at a free flowing outlet of a pipe

93. A hydraulic jump is likely to occur where

    a)   Flow drops from a sewer into a basin

    b)   The velocity of flow is suddenly increased

    c)   The flow depth is greater than the critical depth

    d)   Wastewater moving at a high velocity at shallow depth strikes wastewater having a substantial depth

94. Inverted siphons are used to

    a)   Prime pumps in a pumping station

    b)   Prevent flooding of pumping stations

    c)   Cross depressions or small water courses

    d)   Connect an individual house when elevations are low on the property owner's side

95. The hydraulic head losses associated with transitions are

    a)   Static losses

    b)   Hydrostatic losses

    c)   Velocity reductions

    d)   Friction and conversion losses

96. If you noticed that raw wastewater had developed a blackish color with an unpleasant sour odor and by testing you found the pH was down from a normal 7.2 to a pH value of 6.1, this would indicate

a) High infiltration

b) High oxygen content

c) Anaerobic decomposition (septic wastewater)

d) Possible discharge of acid waste into the sewer system

97. Crown corrosion typically results from

a) Two or more gases when they mix together

b) Hydrogen sulfide in the presence of moisture forms an acid that deteriorates the top of the sewer pipes

c) The excessive amount of nitrates in the wastewater

d) None of the above

98. Odors in the collection systems may be caused by

a) Aerobic bacteria

b) Anaerobic bacteria

c) Psychrophilic organisms

d) Inorganic chemicals reacting with inert solids

99. The difference between static head and dynamic head is

(1) Static head is the difference between the water surface elevation on the suction side of a pump and water surface on the discharge side of a pump.

(2) Static head is the difference between the elevation at the bottom of the wet well and the elevation from where wastewater is discharged.

(3) Dynamic head is the static head plus one-half the pipe coefficient.

(4) Dynamic head is the static head plus the friction or energy losses that result from liquid flowing through the pipes, valves, and fittings in the lift station.

a) 1 and 2 above

b) 1 and 3 above

c) 1 and 4 above

d) 4 and 3 above

100.   A manhole shelf is

    a)   A portable workbench that can fit down a manhole

    b)   The working areas on either side of the invert

    c)   The lip that retains the manhole cover

    d)   None of the above

101.   All sewers should be designed and constructed with hydraulic slopes sufficient to give mean velocities when flowing full or half full of not less than

    a)   1.0 mph (1.6 km/h)

    b)   2.0 ft/min (10.2 mm/s)

    c)   2.0 ft/sec (0.61 m/s)

    d)   3.1 ft/sec (0.96 m/s)

102.   A check valve is best defined as

    a)   A sluice valve

    b)   A valve to prevent backflow

    c)   A globe valve

    d)   A unidirectional valve

103.   A drop manhole

    a)   Accommodates lines at various grades

    b)   Dissipates velocity

    c)   Reduces splashing

    d)   All of the above

104.   The decomposition and decay of organic material in a sewer system containing no "free" or dissolved oxygen is called

    a)   Aerobic

    b)   Anaerobic

    c)   Ambient

    d)   Auxiliary

Answers for Section 1—Background Knowledge

1.	d	24.	d	47.	c	70.	c	93.	d
2.	b	25.	d	48.	b	71.	b	94.	c
3.	c	26.	d	49.	c	72.	b	95.	d
4.	c	27.	b	50.	b	73.	d	96.	c
5.	d	28.	d	51.	c	74.	b	97.	b
6.	b	29.	b	52.	b	75.	a	98.	b
7.	c	30.	b	53.	c	76.	b	99.	c
8.	b	31.	d	54.	c	77.	c	100.	b
9.	a	32.	d	55.	d	78.	c	101.	c
10.	d	33.	c	56.	c	79.	b	102.	b
11.	d	34.	c	57.	d	80.	d	103.	d
12.	c	35.	d	58.	c	81.	b	104.	b
13.	d	36.	c	59.	d	82.	c		
14.	b	37.	a	60.	b	83.	d		
15.	c	38.	a	61.	c	84.	a		
16.	c	39.	b	62.	c	85.	c		
17.	c	40.	d	63.	d	86.	b		
18.	c	41.	a	64.	d	87.	b		
19.	b	42.	c	65.	c	88.	d		
20.	c	43.	d	66.	d	89.	c		
21.	b	44.	d	67.	a	90.	d		
22.	b	45.	d	68.	c	91.	b		
23.	c	46.	c	69.	d	92.	d		

2 SUPPORT SYSTEMS

Pumping/lift stations are a vital component of almost every wastewater collection system. Sometimes referred to as *lift stations*, pumping stations are used to raise wastewater from a lower elevation to a higher elevation. There are several distinct types of pumping/lift stations covered in the following questions.

1. Of the following devices, which is used for measuring pressurized wastewater flow?

    a) Parshall flume
    b) Metering valve
    c) Surface level indicator
    d) Magnetic flow meter

2. How much air pressure will be required to overcome the water pressure in a pressure-sensing controller, for every foot (meter) of water rise in a wet well?

    a) 0.43 psi/ft (9.79 kPa/m)
    b) 0.87 psi/ft (19.59 kPa/m)
    c) 7.48 psi/ft (187.9 kPa/m)
    d) 0.34 psi/ft (188.7 kPa/m)

3. Which of the following is *not* a type or component of pump control systems used at wastewater lift stations?

    a) Bubbler tubes with backpressure switches
    b) Floats, counterweights, pulleys, and mercury switches
    c) Low-voltage electrodes
    d) Bourden nozzles

4. A wastewater pumping station is designed to

    a) Lift the liquid and not the solids
    b) Lift wastewater
    c) Increase the rate of flow
    d) Store tools and equipment

5.   Manhole covers are typically heavy

   a)   To prevent vandalism
   b)   For public safety
   c)   To support vehicular traffic
   d)   All of the above

6.   The difference between the elevations of the water surfaces in a wet well and at the pump discharge is

   a)   The dynamic head
   b)   The static head
   c)   The velocity head
   d)   The drawdown

7.   A bubbler tube would most likely be found in

   a)   A sewer cleaning truck
   b)   A manhole with little flow
   c)   A pumping station dry well
   d)   A pumping station wet well

8.   Corrosion of metal in lift stations can be controlled by using

   a)   Chemicals
   b)   Heating the station
   c)   Tar paper
   d)   Sacrificial metals

9.   Manholes are typically located at

   a)   Each change in direction of the sewer
   b)   Each change in grade
   c)   Each change in the size of pipe
   d)   All of the above

10.   Telemetry is most commonly associated with

   a)   Measuring flow
   b)   Measuring oxygen
   c)   Communication in an alarm system
   d)   None of the above

11. Grit chambers are designed to remove what from wastewater?

    a) Settleable organic solids
    b) All organic matter
    c) Floating material
    d) Inorganic material such as sand and cinders

12. A wastewater treatment process would *not* be adversely affected by

    a) Excessive amounts of septage
    b) Slug metal loads
    c) High fluctuations in flow
    d) Flows with a pH of 7

13. A wet well is a/an

    a) Ejector tank
    b) Septic tank
    c) Holding tank
    d) Stormwater outlet

14. Float and electrode switches should be checked at least once a week to see that

    a) Motor speed increases gradually
    b) Floatable solids are floating
    c) The switches change the direction of flow
    d) Controls respond to changing water levels

15. The type of flow meter most commonly found in pumping stations is the

    a) Venturi meter
    b) Magnetic flow meter
    c) Parshall flume
    d) All of the above

16. The indicator organisms typically used to determine the pollution of water are

    a) Algae
    b) Fungi
    c) Coliforms
    d) Staphylococcus

17. Lift stations may be required in wastewater collection systems when

   a) Pressure flow is required
   b) The topography is poor for gravity flow
   c) Additional wastewater velocity is required
   d) Additional wastewater flowrates are required

18. The rates of flow through a Parshall flume are determined by measurement of

   a) The wastewater velocity
   b) The flume configuration
   c) The wastewater depth
   d) The wastewater area

19. An operations and maintenance (O&M) manual for a pumping station should include

   a) Agency budget
   b) Safety instructions
   c) Equipment data sheets
   d) Both b and c

20. At what location would you most likely find an air release valve?

   a) At a flushing connection
   b) At a deep manhole on a gravity line
   c) At a high elevation on a force main line
   d) At a shallow manhole on a gravity line

21. The frequency of lift station visits depends on

   a) Number of stations in the community
   b) Condition of equipment
   c) Type of wastewater being conveyed
   d) All of the above

22. Problems that may develop in a wet well that is too large include

   a) Solids deposition
   b) Grease buildup
   c) Septic conditions and odors
   d) All of the above

23. Wastewater collection system operators should be given an opportunity to review the prints and specifications of a new lift station before the award of a construction contract so they can

    a) See if the design engineers know what they are talking about

    b) Possibly prevent future problems by working with the engineers

    c) Begin learning now how to become competent engineers

    d) None of the above

24. Pumping station operators need to be highly qualified to operate and maintain stations so that

    a) Continuous operation is ensured

    b) The municipality's investment is protected

    c) The complexity of the equipment can be understood

    d) All of the above

25. Flow measurement charts are

    a) Useful for comparison to detect inflow

    b) Read immediately

    c) Only used for wastewater treatment facilities

    d) Only used to determine chemical feed rates

26. What information must be included on a warning tag attached to a switch that has been locked out?

    a) Time to unlock the switch

    b) Directions for removing the tag

    c) Signature of person who locked out the switch and who is the only person authorized to remove the tag

    d) All of the above

27. Grease may tend to accumulate on the walls and equipment in a wastewater pumping station wet well. This must be

    a) Dissolved and allowed to enter the treatment process

    b) Encouraged as the grease will react with the detergents in the wastewater and reduce the foaming in the wastewater treatment process

    c) Prevented, if possible, because accumulations often break loose and large pieces may clog the pump suction line or jam the pump control floats

    d) Increased because a coat of grease will protect metal pipes and equipment from rusting

28. The rate of flow in a force main leaving a wastewater pumping station can be measured by a

   a) Disk meter
   b) Positive displacement factor
   c) Calibration meter
   d) Venturi meter

29. Assume that one of your wet wells relies on a bubbler tube system to control the liquid level. If the air line develops a leak above the liquid level, what will this most likely do to your alarms and at what elevation (high or low) do you think you would find the wet well if you were not able to get to the station for 30 minutes and the flow was high?

   a) Make high water alarm sound—high wet well elevation
   b) Make high water alarm sound—low wet well elevation
   c) Make low water alarm sound—low wet well elevation
   d) Make low water alarm sound—high wet well elevation

30. The body of a 6-in. gate valve has "200 WOG" inscribed on it. This means

   a) The valve is not suited for wastewater service
   b) The catalog number is 200 and the valve can be used in "Waste or Grit"
   c) The manufacturer's code number
   d) The valve is rated for service at 200 psi for oil, gas, or cold water

31. Sodium hypochlorite is

   a) The decomposition product formed when sodium sulfate is used to neutralize a chlorine leak
   b) The salt that is formed when hydrochloric acid is neutralized by sodium hydroxide
   c) A highly insoluble precipitate that forms a scale when saltwater is chlorinated
   d) A compound that can be purchased in liquid solution and can be used for disinfection

32. What is a backflow preventer?

   a) A device to prevent backsiphonage
   b) A device to control water hammer in pressure pipes
   c) An automatic flushing device for instrument purge lines
   d) A device to ensure unidirectional flow in metering runs

33. To reduce the number of mysterious changes in station operation such as pumps turned off, control system changes, and station doors or gates unlocked, pumping stations should have

    a) A 24-hour watch on the station
    b) A sign-in log and established, documented standard operating procedures (SOPs)
    c) Only one person with a key to the station
    d) All of the above

34. What is a cause of cavitation in a pump?

    a) Improper lubrication
    b) Packing gland too tight
    c) Elevated thrust-bearing temperature
    d) Improper design and/or head conditions under which the pump is operated

35. Circumstances that influence the frequency of pumping station visits in a typical collection system may include

    a) Monies allocated in the fiscal budget
    b) Jobs that are on the agenda for the day or week
    c) The number of stations in the system
    d) All of the above

36. Excessive quantities of heavy metals should be excluded from the collection system because

    a) They will damage equipment
    b) They may affect the treatment process
    c) They may eventually enhance the receiving water quality
    d) Both b and c

37. When operating a small wastewater pumping station with two identical pumps, it is best to adjust the controls so that

    a) The pumps turn on together
    b) The pumps alternate in operation
    c) One pump operates continuously, and the other pump is held in reserve until the first one is worn out
    d) The pumps operate only when the wastewater has reached a level at which it would overflow the bypass

38. Maintenance of couplings between the driving and driven elements includes

  a) Keeping proper curvature
  b) Maintaining proper alignment, even with flexible couplings
  c) Draining old oil in fast couplings
  d) Keeping the electrodes greased

39. A *pneumatic ejector* is best defined as

  a) A portable air-compressor eductor
  b) An air tool used to help loosen tight bolts
  c) Compressed air pressurizing a vessel that forces wastewater out through check valves
  d) None of the above

Answers for Section 2—Support Systems

1.	d	**24.**	d
2.	a	**25.**	a
3.	d	**26.**	c
4.	b	**27.**	c
5.	d	**28.**	d
6.	b	**29.**	d
7.	d	**30.**	d
8.	d	**31.**	d
9.	d	**32.**	a
10.	c	**33.**	b
11.	d	**34.**	d
12.	d	**35.**	d
13.	c	**36.**	b
14.	d	**37.**	b
15.	d	**38.**	b
16.	c	**39.**	c
17.	b		
18.	c		
19.	d		
20.	c		
21.	d		
22.	d		
23.	b		

3 OPERATION AND MAINTENANCE

The effective management and operation and maintenance of a wastewater collection system are essential for successful continuous and efficient operation. Neglect and inadequate maintenance can lead to a wastewater collection system's disrepair and failure, potentially costing the utility thousands of dollars in emergency funding.

Because many communities and agencies have made significant capital investments in constructing their wastewater collection systems, highly qualified and certified operators are in great demand to protect these investments. The following questions cover operational and maintenance issues.

1. When cleaning stoppages between property lines and the main line sewer, the access is called a/an

   a) Manhole
   b) Invert
   c) Cleanout
   d) Channel

2. Which of the following would *not* cause stoppages in sewers?

   a) Adverse hydraulic conditions
   b) Grease
   c) High velocities
   d) Roots

3. Metal detectors may be used to locate

   a) Asbestos-cement pipe
   b) Vitrified clay pipe
   c) Buried manhole covers
   d) All of the above

4. Normal hydraulic cleaning of a sewer line removes

   a) Gases
   b) Grit
   c) Heavy debris
   d) All of the above

5. Efforts to eliminate infiltration and inflow in a collection system are made primarily to

   a) Reduce hydraulic loads
   b) Reduce dilution of wastewater
   c) Prevent debris from entering into pipes
   d) Provide for a higher groundwater table

6. A bell of a pipe that is chipped during installation should not be left in place because it may cause _____.

   a) Infiltration
   b) Exfiltration
   c) Root intrusion
   d) All of the above

7. The atmosphere in a sewer system may be

   a) Tested with the proper equipment
   b) Corrosive to metal
   c) Perfectly okay to breathe
   d) All of the above

8. When nearby residents complain of objectionable odors emanating from a pumping station, you should

   a) Tell the residents that you have no control over odors
   b) Use copious amounts of lime to raise the pH of the septic wastewater
   c) Determine the cause of the odors and take corrective action
   d) Use commercial masking agents

9. When working in a manhole and a rumbling sound is heard, what is most likely happening?

   a) Light traffic in the distance
   b) Increase in flow from nearby lift station
   c) Rodents in the system have been disturbed
   d) None of the above

10. Of the following chemicals listed, which is/are of benefit for odor control at pumping stations or in the collection system?

   a) Chlorine
   b) Hydrochloric acid
   c) Hydrogen peroxide
   d) Both a and c

11. The location of the clearing tool must be known at all times when using a power-rodding machine so

   a) There is enough rod left over if it breaks
   b) You know where to place the debris trap
   c) You will know where to place the leader tool if the rod becomes coiled
   d) You will know where to dig to recover the tool if it becomes stuck

12. In which direction would you place the nozzle in the sewer when using high-velocity cleaning machines?

   a) Upstream in the sewer against the flow
   b) Downstream in the sewer with the flow
   c) Does not matter
   d) Depends on the velocity

13. Suppose an 8-in. (200-mm) diameter sewer line is backing up, the problem being caused by an excessive grease buildup. Which of the following pieces of equipment could *not* be used to alleviate the problem?

   a) Bucket machine
   b) Chemicals or bacteria
   c) Ball and line
   d) Power rodder

14. Which of the following sewer rodding tools can be used when roots are a problem?

    a) Square stock corkscrew
    b) Auger
    c) Rootsaw
    d) All of the above

15. Which of the listed methods cleans sewer lines hydraulically?

    a) Chemical addition
    b) High-velocity jetting machine
    c) Power bucket machine
    d) Power-rodding machine

16. What does the term "porcupine" refer to in the context of a collection system?

    a) A bucket machine with spikes
    b) A chemical used to dissolve grease
    c) A cleaning tool used to scour pipes
    d) None of the above

17. After using rodding equipment to clear a blocked sewer, you should then

    a) Allow it to rest
    b) Clear it with a bucket machine
    c) Check it for proper infiltration
    d) Clean it hydraulically to restore full capacity

18. In wastewater collection systems, television inspection is used to

    a) Determine what joints have to be sealed
    b) Regulate flow
    c) Determine where sources of infiltration originate
    d) Both a and c

19. What is the main purpose of a manhole?

    a) A sampling point for wastewater
    b) For water storage in case of a blockage
    c) For access to inspect or clean lines
    d) To trap rodents

20. Which one of the following tools is *not* used in rodding a sewer line?

    a) Porcupine
    b) Lag screw pull
    c) Rootsaw
    d) Sand corkscrew

21. One reason for knowing the exact location of a cleaning tool as you clean a line is so

    a) The proper charges can be made to homeowners
    b) The slope of the sewer can be computed and recorded
    c) The valves inside the sewer can be reopened after the line is cleaned
    d) You can record where line stoppages have occurred

22. To be most effective in reducing or eliminating odors, chlorine must be added

    a) At the source of the odor problem
    b) Upstream of the problem area
    c) Downstream of the problem area
    d) None of the above

23. The major reasons for equipment maintenance are

    a) Reliability
    b) Cost-effectiveness
    c) Safety
    d) All of the above

24. Solids that settle in a sewer line will most likely result in the generation of

    a) Carbon monoxide
    b) Hydrogen sulfide
    c) Oxygen
    d) Methanol

25. The part(s) of a pipe that is/are most vulnerable to corrosion from septic wastewater in a gravity collection system is/are the

    a) Bell
    b) Crown
    c) Flange and bolts
    d) Invert

26. Hydrogen sulfide may cause the following conditions:

    a) Rotten egg odor
    b) Excessive pH values
    c) Corrosion of sewers, structures, and equipment
    d) Both a and c

27. Odors will typically be detected in sewers that are

    a) Experiencing substantial amounts of infiltration
    b) Hydraulically underloaded
    c) Relatively short in total mileage piped
    d) None of the above

28. Of the following, which is the *most* important for a good preventive maintenance program?

    a) Adequate budget
    b) Sufficient number of personnel
    c) Good records
    d) Good equipment

29. The application of 10 mg/L of chlorine to wastewater at a pumping station

    a) Will cause problems in the operation of the water resource recovery facility
    b) Is not permitted
    c) Should only be done during rainstorms
    d) Is used for odor control

30. A good collection system maintenance program includes which of the following?

    a) A regular inspection schedule
    b) A thorough knowledge of collection systems
    c) An adequate recordkeeping system with sewer line location maps
    d) All of the above

31. The purpose of a maintenance schedule is

    a) To train new personnel
    b) For equipment replacement
    c) To spend the monies that were appropriated
    d) To help prevent problems from developing

32. Roots may be temporarily removed from sewers by the use of

a) Power rodders

b) Addition of a diluted chlorine solution

c) Pumping

d) Flushing

33. Hydrogen sulfide can be found under which of the following conditions?

a) At most house services

b) A fast-flowing trunk line

c) Sluggish lines that are coated on the inside with slimes

d) Lines that service an industrial waste complex with a high pH

34. When we talk about good housekeeping around a pumping station, we mean

a) Reading magazines

b) Keeping coffee fresh and warm

c) Not hosing down all spills immediately

d) Providing a proper place for tools and equipment

35. When placing a manhole hook under a sewer manhole cover, the operator should pull

a) Upward

b) With his/her back to the traffic

c) Does not matter

d) With his/her back in line with the center of the road

36. In the event that a forced draft ventilation fan in a pumping station fails, it must be fixed immediately because

a) The pumps will overheat

b) Bacterial contamination could occur

c) Exclusive gases could accumulate

d) All of the above

37. When operating a power-rodding machine, it is extremely important that the operator

a) Extends the rod to the next manhole before starting to rotate the tool

b) Has a district or maintenance supervisor present at all times

c) Has the rod moving in or out of the sewer whenever it is rotating

d) Avoids rodding in a sewer unless there is a full flow in the sewer line

38. One practical method of removing roots in a collection system would be to

    a)   Excavate the pipe

    b)   Use an axe or chainsaw

    c)   Use a flexible rod and cutter

    d)   Apply heavy doses of chemicals

39. How can members of a balling crew most effectively communicate with each other?

    a)   Writing

    b)   Telegraph

    c)   Hand signals

    d)   By shouting down the sewer

40. When using a hydraulic sewer-cleaning method, care must be taken

    a)   To prevent any air gap from occurring

    b)   To always plug the downstream manhole

    c)   Not to cause flooding in homes and basements

    d)   To throttle flows from fire hydrants by using the hydrant valves

41. Of the following items, which would be of *least* importance when filling out the daily report for a high-velocity sewer-cleaning crew?

    a)   Total distance of sewers cleaned that day

    b)   Sizes of sewers cleaned that day

    c)   The number of sewer rods broken each day

    d)   Specific identification of the sewers cleaned

42. The main objective for the operation and maintenance of a wastewater collection system is

    a)   Keeping an eye out for vandals

    b)   Keeping organic solids out of the effluent

    c)   Keeping all industrial discharges from entering the collection system

    d)   None of the above

43. Blockages can be positively identified, and the location pinpointed by

    a)   Dye tests

    b)   Smoke tests

    c)   Trees growing over the sewer

    d)   Closed-circuit television inspection

44. Flushing of a sewer line would *best* remove

  a) Grit
  b) Roots
  c) Heavy debris
  d) None of the above

45. The proper care of hand tools requires that they

  a) Be cleaned after use
  b) Be kept in good condition
  c) Be used only for their intended use
  d) All of the above

46. A manhole should be ventilated

  a) During hot weather
  b) During low-flow periods
  c) When methane gas is present
  d) When more than 20% oxygen is present

47. When a sewer is being rodded or jetted to clear a blockage, the work is typically started from the

  a) Wet manhole
  b) Dry manhole
  c) Chimney between manholes
  d) Flooded manhole if the wastewater is not too deep

48. When a mechanical blower is used to ventilate a manhole

  a) The blower should discharge into the manhole
  b) The suction of the blower should exhaust air from the manhole
  c) Both suction and discharge lines should be in the manhole
  d) None of the above

49. Hydraulic pipe cutters are typically used to cut

  a) Vitrified clay pipe
  b) Asbestos cement pipe
  c) Reinforced concrete pipe
  d) Cast iron pipe

50. Smoke testing of sewers

    a)   Is a process for determining if sewer gas is present in a manhole

    b)   Can be used to detect downspout connections to a sewer

    c)   May be used to measure the quantity of infiltration when the groundwater table is low

    d)   Both b and c

51. Which of the following pipe materials is most commonly corroded by sewer gases?

    a)   Vitrified clay

    b)   Polyvinyl chloride (PVC)

    c)   Reinforced concrete

    d)   Ductile iron

52. When filling a hydraulic cleaner with water from a fire hydrant

    a)   You should only fill the tank three-quarters full

    b)   Make sure the fire hydrant is open halfway

    c)   Be certain the fill hose extends well into the tank

    d)   Be sure there is an air gap between the fill hose end and the water level

53. A rodding machine sits 30 ft (9.1 m) away from a manhole and 145 ft (44.2 m) of rod is taken off the reel. Assuming that the manhole is 10 ft (3.1 m) deep, how far into the sewer line is the end of the rod?

    a)   105 ft (32.0 m)

    b)   115 ft (35.1 m)

    c)   125 ft (38.1 m)

    d)   155 ft (47.2 m)

54. The rate of travel of a sewer ball through the line must be controlled to

    a)   Move debris

    b)   Remove slime

    c)   Maintain a constant speed

    d)   Prevent the wastewater from surcharging upstream

55. A bucket machine may be used to clean a sewer that is

  a) Loaded with grease
  b) Plugged with roots
  c) Composed of vitrified clay and full of large rocks
  d) On a flat grade

56. The most effective method used to check for illegal connections to a wastewater collection system is

  a) Dye testing
  b) Smoke testing
  c) Flow measurement
  d) A telephone survey

57. Infiltration, or the entry of stormwater and groundwater into the collection system, causes collection systems and water resource recovery facilities to increase in size. Which of the problems listed below causes the greatest volume of infiltration to the system?

  a) Residential sump pump discharges tied into the sanitary sewer
  b) Faulty house sewers
  c) Faulty joints in trunk sewers
  d) Crushed sections of sewer lines

58. When performing repairs to a sewer line under a public road, which of the following is/are essential step(s)?

  a) Inform the public in advance if the work is to be of a major scope
  b) Clearly and properly install all barricades and safety devices at the construction site
  c) Contact the fire and police departments to inform them of where and when construction will occur
  d) All of the above

59. When using a sewer ball to clean a sewer, in which direction would the ball be moving?

  a) Upstream
  b) Downstream
  c) It does not matter
  d) Always toward the nearest lift station

60. Roots in sewers causing stoppages can be positively identified as the source of the problem by

    a) Dye tests
    b) Smoke tests
    c) Closed-circuit television inspection
    d) Observing roots in flow downstream, after clearing a stoppage with a power rodder

61. If a sewer line is known to have root problems, you could use a _____ to improve the condition.

    a) Shield
    b) Scooter
    c) Power rodder
    d) Sewer ball

62. The difference between a masking agent and a chemical such as potassium permanganate is

    a) The masking agent is a permanent solution and potassium permanganate is not
    b) Potassium permanganate is very costly and masking agents can easily be made
    c) Masking agents cover up and potassium permanganate chemically oxidizes the odor-causing material
    d) None of the above

63. The flow velocity in a sewer may be measured by a

    a) Weir
    b) Parshall flume
    c) Venturi meter
    d) None of the above

64. Catch basin inlets to stormwater sewers have been used primarily to

    a) Settle out sand and grit from the stormwater flow
    b) Prevent the dissemination of odors from the sewer
    c) Act as an equalizer of the flow to avoid overtaxing the capacity of the sewer
    d) Prevent floating objects from entering the storm sewer obstructing the flow in the sewer

65. The purpose of smoke testing is to

  a) Eliminate insects and rodents from the collection system
  b) Measure the volume of air in a sewer line from manhole to manhole
  c) Determine which way the water will flow
  d) Locate inflow sources

66. High velocity cleaners use _____ to remove debris from the sewers.

  a) Centrifugal force
  b) Cutting and scraping action
  c) High water pressure
  d) All of the above

67. Problems that may arise when a check valve leaks include

  a) The check valve seat may wear prematurely
  b) The pump can spin backward
  c) Electrical costs will be higher because of repumping of previously pumped waste-water
  d) All of the above

68. Which of the following equipment is *not* necessary when conducting a closed-circuit television examination of a sewer?

  a) Picture monitor
  b) AM-FM radio
  c) Cable-pulling winch
  d) Communication system

69. Odors in manholes at the end of force mains are typically caused by

  a) Turbulence
  b) Aerobic bacteria
  c) Anaerobic conditions
  d) Both a and c

70. Sources of infiltration can *best* be located by

  a) Smoke tests
  b) Visual tests
  c) Flow metering
  d) Television inspection

71. When placing a weir in an open channel, the measurement of the head should be made upstream of the weir. To avoid surface drawdown that occurs from the crest, the measurement location should be at least how far away from the weir?

    a)   At the weir crest

    b)   Twice the head

    c)   Four times the head

    d)   Just behind the crest

72. The best method to initially use to retrieve a broken section of sewer rod that has snapped inside an 8-in. (200-mm) diameter line, assuming that the broken rod comes from a power rodder and that it *cannot* be retrieved by hand, is

    a)   Look at your footage meter on the rodder, mark out the distance on the street, and excavate

    b)   Try and push the rod down to the next manhole

    c)   Use a rod retriever

    d)   None of the above

73. Exfiltration should be controlled to prevent

    a)   Possible groundwater contamination

    b)   A depletion of wastewater that is needed to operate a treatment facility properly

    c)   Backfill from washing away

    d)   Both a and c

74. To control hydrogen sulfide and odors in a 12-in. (300-mm) diameter sewer, the chlorine dosage must be 10 mg/L. When the flow is 0.37 mgd (1400 m^3/d), the feed rate is

    a)   15.4 lb/d (7.0 kg/d)

    b)   30.8 lb/d (14 kg/d)

    c)   154 lb/d (70 kg/d)

    d)   308 lb/d (140 kg/d)

75. Closed-circuit television units with videotape can help to evaluate the condition of a wastewater collection system by

    a)   Providing live inspection during use

    b)   Giving the exact location of a problem

    c)   Studying the effectiveness of a cleaning technique

    d)   All of the above

76. When measuring flow velocities between manholes using a floating object in a sewer flowing half full, the velocity of that float will be

    a) The average velocity of the wastewater

    b) 10 to 15% slower than the average velocity of the wastewater

    c) 10 to 15% faster than the average velocity of the wastewater

    d) 50% of the average velocity of the wastewater

77. Some of the causes of physical injuries in manholes include

    a) Slips and falls

    b) Failure to use proper lockout/tagout procedures

    c) Infection and disease

    d) All of the above

78. The main reason for conducting a dye test instead of a smoke test is because

    a) It is less expensive than smoke

    b) The service connection could have a dip in it or the trap may be full of water

    c) It takes less time to do the test

    d) None of the above

79. Some ways to determine if a certain house is connected to a sanitary sewer include using

    a) Dye testing

    b) Some tests

    c) Closed-circuit television inspection

    d) Both a and b

80. Which of the following would *not* be a reason for prechlorination, either at the water resource recovery facility inlet or somewhere in the collection system?

    a) To improve grit removal

    b) To help control corrosion in long force mains

    c) To control odors at the headworks

    d) To accomplish at least partial disinfection during overflow emergencies

81. Obstructions in sewers can *best* be cleared by

a) Very small blasting charges
b) A power rodder with an auger head
c) Lamping between manholes
d) Using a hose from the nearest fire hydrant

82. You have a new gasoline-powered diaphragm-type pump for dewatering at sewer repair sites. You are using heavy-duty, spiral-wire-reinforced suction hose. Which of the following is a practical limit of height to set the pump above the water level in the ditch?

a) 3 to 8 ft (0.9 to 2.4 m)—if you try to pump with more than 8 ft (2.4 m) of suction lift, the plies of the pump diaphragm will rupture
b) 20 to 25 ft (6.1 to 7.6 m)—this is approaching the suction lift limits because of barometric pressure
c) 40 ft (12.2 m)—safety regulations limit net suction lift to 40 ft (12.2 m) or less for gasoline-powered field pumps
d) 65 ft (19.8 m)—the suction hose would collapse with more than 65 ft (19.8 m) of negative water column

83. It is recognized that the soundest mechanical equipment maintenance program, in addition to proper lubrication, includes

a) No other key features
b) Continuous use of machinery until it breaks down, after which it is replaced
c) Periodic inspection of each part of all mechanical equipment and repair or replacement of their parts, when needed, before failure
d) Replacement of machine parts at regular intervals

84. In preparing an effective schedule for maintaining a wastewater collection system, which of the following factors should be considered?

a) Checking system records
b) "Benchmarking" against high-performing systems
c) Past experience of system performance
d) All of the above

85. A person may gain knowledge or become familiar with a collection system by

a) Training and seminars
b) Being patient—in time you will learn
c) Learning from experienced fellow workers
d) Both a and c

86. Which of the following is the correct procedure for opening and closing valves on high-pressure lines?

a) They should be opened and closed very slowly

b) They should be opened and closed as quickly as possible

c) No special procedures are required

d) All valves should be closed by turning in a clockwise direction and opened by turning counter clockwise

87. Corrosion control may be accomplished by which of the following methods?

a) Increasing the oxygen content

b) Increasing the carbon dioxide content

c) Increasing the hydrogen ion concentration

d) Decreasing the volume of flow from the pumping stations

88. Typically, hydrogen sulfide will be found under what conditions?

a) Fast-flowing trunk line

b) A line that services an industrial waste complex with a high pH

c) A sluggish line that is coated on the inside with algal slime

d) At most house service connections

89. A pig would most likely be used in a

a) Gravity sewer

b) Storm sewer

c) Force main

d) House or building sewer

Answers for Section 3—Operation and Maintenance

1.	c	24.	b	47.	b	70.	d
2.	c	25.	b	48.	a	71.	c
3.	c	26.	d	49.	d	72.	c
4.	b	27.	b	50.	b	73.	d
5.	a	28.	c	51.	c	74.	b
6.	d	29.	d	52.	d	75.	d
7.	d	30.	d	53.	a	76.	c
8.	c	31.	d	54.	d	77.	d
9.	b	32.	a	55.	a	78.	b
10.	d	33.	c	56.	a	79.	d
11.	d	34.	d	57.	c	80.	a
12.	a	35.	d	58.	d	81.	b
13.	a	36.	c	59.	b	82.	b
14.	c	37.	c	60.	c	83.	c
15.	b	38.	c	61.	c	84.	d
16.	c	39.	c	62.	c	85.	d
17.	d	40.	c	63.	d	86.	a
18.	d	41.	c	64.	a	87.	a
19.	c	42.	d	65.	d	88.	c
20.	d	43.	d	66.	c	89.	c
21.	d	44.	a	67.	d		
22.	a	45.	d	68.	b		
23.	d	46.	c	69.	d		

4 SUPERVISION AND MANAGEMENT

To maintain an efficient, effective, and safe wastewater collection system, sound system management and operation and maintenance are critical. Without sound management, structure, and discipline, any programs that are currently in place will most likely be ineffective and inefficient.

This section focuses on practical, real-world management questions for personnel working at various levels within the collection system.

1. Which of the following elements should be contained in an effective sewer use ordinance?

   a) An organizational chart
   b) Prohibited wastes
   c) A preventive maintenance schedule
   d) Sewer design criteria

2. Upon opening a sewer manhole, you detect a very strong gasoline odor. Which of the following should you do first?

   a) Close the lid carefully and wait for it to pass
   b) Start looking for the source of the problem
   c) Call your supervisor immediately; if he/she is not available, notify the fire and police departments
   d) Make a note in your daily report so that conditions can be compared at the next manhole inspection

3. During the winter months when you are down in a manhole, how far away should your surface partners be?

   a) One at the rim and another in the truck keeping warm
   b) Both in the truck, but available on the radio
   c) Both at the rim
   d) None of the above

4. The supervisor of a repair crew should

   a) Keep an inventory of all tools
   b) Keep a log of time needed to do the repair
   c) Never leave the job site
   d) Both a and b

5. How should a supervisor deal with the constant neglect of safety rules by an employee?

   a) Make the employee sign a "hold harmless" agreement
   b) Joke about his/her actions to keep a positive team-based atmosphere
   c) Initiate immediate disciplinary action
   d) Relax safety rules for everyone in fairness to all employees

6. When a complaint of a plugged sewer is received, your first action would be to

   a) Decide whether you are too busy to deal with the problem thoroughly and, if so, ask the caller to call back tomorrow
   b) Repair it when you have the time
   c) Go to the site to confirm the problem
   d) All of the above

7. Your supervisor asks if you have any ideas that will help prevent the opening of manholes by vandals; your best suggestion might be to

   a) Spot weld the manhole covers
   b) Have a crew ride around and possibly catch vandals in the act
   c) Attach an alarm system that will sound when the manhole is opened
   d) None of the above

8. When evaluating the application of chemicals to alleviate a problem in a sewer, remember that the chemicals

   a) Correct the source
   b) Are a permanent solution
   c) Perform the same in the field as in the laboratory
   d) Require consideration of the workers and public

9. Your municipality seeks your advice on an ordinance regarding the use of the collection system. Of the following choices, which statement would not be included as a provision in the ordinance?

    a) The discharge of flammable chemicals is prohibited
    b) The pretreatment of strong industrial wastes is required
    c) Flow from domestic house connections must be strictly controlled
    d) Rainwater downspouts must be connected into a "separate" system

10. Good public relations benefit

    a) The public
    b) Employees
    c) Government agencies
    d) All of the above

11. Staffing requirements for a collection system should be based on

    a) Work load
    b) Survey of the staff's opinion
    c) Historical practices
    d) Both b and c

12. A reasonably accurate operational report of the collection system is

    a) Of no value to anyone
    b) Of the greatest value to the operator and municipal officials
    c) Beneficial only to state officials
    d) Of value only to those working in the system

13. Easement agreements should contain a right of access for

    a) Repairs
    b) Inspection
    c) Maintenance
    d) All of the above

14. The responsibility of providing your agency with a good recordkeeping system lies with

    a) The foreman
    b) The supervisor
    c) The field workers
    d) None of the above

15. The requirements for tools and equipment necessary to properly operate a collection system depend on

    a) Objectives
    b) Type of equipment to be worked on
    c) Size of the collection system
    d) All of the above

16. To enhance public relations, collection systems operators should strive to present

    a) A neat appearance
    b) The proper attitude
    c) A good first impression
    d) All of the above

17. A telephone log of a complaint should contain which of the following?

    a) Name of caller
    b) Time call was received
    c) Nature of complaint
    d) All of the above

18. If a local tannery suddenly starts discharging large quantities of buffing dust that interfere with the satisfactory operation of the collection system, the operator should first

    a) Notify the state regulatory agency so that they can order the tannery to stop discharging the dust
    b) Call the local police and have them order the tannery to keep the dust out of the sewer in accordance with the city ordinance
    c) Telephone the tannery and tell them you will notify the police if they do not stop immediately
    d) Contact the tannery, explain your operating problems, and offer to cooperate with them in working out a satisfactory solution for disposing of the dust

19. A new employee comes to you with a suggestion for changing the method of performing a routine job. As a foreman, you should

    a) Tell this employee that routine procedures are standardized and cannot be changed

    b) Not consider the suggestion because this person is not familiar with the work

    c) Tell the person to try out the new method unofficially to see if it works

    d) Discuss the suggestion with the employee and determine its value

20. If the pH of the wastewater changed suddenly at the same time of day each week, a probable cause would be

    a) Industrial wastes

    b) Grease traps in local casinos are being bypassed

    c) Scavengers are dumping septic tank sludge into sewers

    d) Both a and b

21. Of the following, it would be poor supervision by a foreman if he/she

    a) Consulted the assistant supervisor on unusual problems

    b) Made it a policy to avoid criticizing an employee in front of his/her co-workers

    c) Allowed a cooling off period of several days before giving one of his/her employees a deserved reprimand

    d) Asked an experienced collection system worker for his/her opinion regarding a special installation project

22. Which of the following would not be a legitimate reason for bypassing your pumping station?

    a) Important major repair

    b) Broken shaft on odor control equipment

    c) Flows in excess of influent line capacity

    d) An extended power outage causing loss of pumping

23. Management can motivate workers by promoting

    a) Good communication

    b) Employee participation in decision-making

    c) Recognition programs for good performance

    d) All of the above

24. Before recommending that action be taken against a subordinate for an alleged infraction of the rules and regulations, you should make absolutely sure that

   a) The charges will be sustained at the hearing
   b) You have all the pertinent information on the incident
   c) Your supervisor will give his/her approval to the recommendation
   d) Fellow employees will give favorable testimony to your side of the case

25. As a foreman, it is necessary for you to know that your operators are learning proper work procedures. One good, practical method to use in determining this in the case of a new employee is to

   a) Assume he/she is performing well if you do not receive any negative feedback
   b) Ask other employees how this person is doing
   c) Assume that if no questions are asked he/she understands his/her duties
   d) Inspect the work that has been assigned to the new employee to give him/her feedback

26. A briefing before manhole entry should discuss which of the following topics?

   a) The job to be done
   b) Problems that might arise
   c) Safety procedures in case of an emergency
   d) All of the above

27. If you were informed that a large quantity of acid waste was accidentally discharged into the sanitary sewer in your city, you could verify this because

   a) The pH of the raw wastewater would rise
   b) The pH of the raw wastewater would drop
   c) The volatile portion of the suspended solids would be greatly increased
   d) The dissolved oxygen level in the raw wastewater would exceed the saturation value

28. The types of records needed to operate and maintain wastewater collection systems include

   a) Costs
   b) Personnel
   c) Equipment maintenance
   d) All of the above

29. Assume that a new hire is assigned to your crew and this worker has very little experience on work performed by your crew. This worker could be best trained by

    a) Having him/her read books on the work
    b) Immediately assigning full duties and allowing him/her to learn by doing
    c) Giving him/her your personal attention until he/she becomes proficient
    d) Assigning him/her to work with an experienced worker who will help him/her along

30. Good public relations are

    a) A bothersome task
    b) A job only for management
    c) Not needed in this business
    d) Essential for continued effective service

31. Before digging in the street to install a new sewer line, you should always

    a) Dig carefully
    b) Know how deep to dig
    c) Notify your supervisor of where you plan to be working
    d) Instruct your equipment operator on the best method to clean his/her machinery

32. One method that generally is helpful in encouraging an employee to derive satisfaction from his/her work is by offering a/an

    a) Opportunity to demonstrate his/her proficiency in various types of work by frequently changing his/her responsibilities
    b) Change in position whenever a need for self-expression is indicated
    c) Realization that job security depends on an unquestionable interest in the job
    d) Understanding of the relationship of his/her job to the work of the whole organization

33. Maps can be used as a maintenance record to

    a) Show trouble spots
    b) Determine size of pipes
    c) Indicate type and age of pipe
    d) All of the above

34. A person can gain knowledge or become familiar with a collection system by

    a)   Obtaining hands-on experience

    b)   Getting education and attending seminars

    c)   Reading books and journals

    d)   All of the above

35. How can organizational charts be useful?

    a)   They ensure everyone will know their job responsibility

    b)   They ensure basic safety concepts are thoroughly understood by all

    c)   They show the chain of command and document overall organizational structure

    d)   They can outline an effective employee training program

36. As a collection system superintendent, you must make sure that someone is primarily responsible for safety supervision. Which of the following approaches would you consider most appropriate?

    a)   Rotate the responsibility to someone new each month so everyone will feel that they are important

    b)   Do it yourself; no one else would have enough authority to enforce safety regulations throughout the entire collection system

    c)   Pick a permanent safety supervisor so that the individual can provide planning, continuity, and follow-up as hazards are noted or reported

    d)   Ask for a volunteer; if you have to appoint a draftee, he/she would not be as conscientious and capable as someone who had expressed an interest in the job

37. If operators working under your supervision are non-cooperative, what question should you ask yourself?

    a)   Am I interested in them?

    b)   Do I really know them?

    c)   Who do they think they are? They should do as they are told!

    d)   Is my leadership effective?

38. A sewer use ordinance

    a)   Should be familiar to all utility employees

    b)   Is designed to keep industries from discharging anything to the sewers

    c)   Indicates who does the monitoring and how

    d)   Both a and b

39. Records are *best* used for

a)   Justifying salary increases

b)   Department use only

c)   Anticipating and identifying operational issues

d)   Showing taxpayers what has been done

40. Industrial waste ordinances typically do not contain specific limits on levels of

a)   Suspended solids

b)   Biological oxygen demand

c)   Dissolved oxygen

d)   pH

41. Which of the following approaches, if used properly by management, is most likely to be effective on a day-to-day basis in affecting worker motivation?

a)   Addition of fringe benefits

b)   Pay increases

c)   Worker recognition

d)   Dictatorial supervision of workers

42. Assume a worker has been injured while on the job. As a responsible supervisor, what steps might you take to ensure that the accident will not be repeated?

a)   Find out how it happened

b)   Correct the cause of the problem

c)   Implement changes to the safety program if indicated by the accident investigation

d)   All of the above

43. What would you do after you have selected a method of solving a collection system problem and you think you have corrected the problem?

a)   File a report of all results

b)   Approach your supervisor and ask for a raise

c)   Check with follow-up inspections at reasonable intervals

d)   Both a and c

44. Some of the advantages of contracting out capital improvements include

    a)   It provides justification for maintaining a larger crew

    b)   There is less maintenance required on equipment that has been installed under a contract

    c)   There is less time available for crews to perform preventive maintenance

    d)   There is more time for crews to perform preventive maintenance

45. You are asked to justify your capital equipment requests to the town council at the annual budget hearing. Of the following, which resources would you call upon as backup to clarify your position?

    a)   Past records

    b)   Equipment life

    c)   Inflation, because the cost to operate a system increases each year

    d)   All of the above

46. Organization charts can be helpful by showing

    a)   Direction of flow throughout the collection system

    b)   The benchmarks by which employees are upgraded

    c)   The structure of an agency

    d)   None of the above

47. Grease and oil concentrations are typically controlled by the city or town

    a)   Sewer user charge

    b)   Sewer use ordinance

    c)   Public health officials

    d)   None of the above

48. A good sewer use ordinance should provide for the following items:

    a)   Require neutralization of acid wastes

    b)   Prohibit any discharge of flammable materials into the sewers

    c)   Provide for the inspection of all sewer line gradients and house connections

    d)   All of the above

49. The *best* way to record the completion of work is through a

   a)   Closed-circuit television
   b)   Polaroid system
   c)   Tape recorder
   d)   Work order system

50. The most important function of recordkeeping is to record

   a)   How much money was budgeted and how much was spent
   b)   The past, to provide a sound basis on which to plan for the future
   c)   What was done and when it was done
   d)   Where we have been and how we got where we are today

51. Disaster planning is

   a)   Having manuals ready so they can be read if a disaster occurs
   b)   Something that, if properly done, will not need to be revised
   c)   Designed to reduce confusion in the event of a disaster
   d)   None of the above

Answers for Section 4—Supervision and Maintenance

1.	b	24.	b	47.	b
2.	c	25.	d	48.	d
3.	c	26.	d	49.	d
4.	d	27.	b	50.	b
5.	c	28.	d	51.	c
6.	c	29.	d		
7.	a	30.	d		
8.	d	31.	b		
9.	c	32.	d		
10.	d	33.	d		
11.	a	34.	d		
12.	b	35.	c		
13.	d	36.	c		
14.	b	37.	d		
15.	d	38.	d		
16.	d	39.	c		
17.	d	40.	c		
18.	d	41.	c		
19.	d	42.	d		
20.	a	43.	d		
21.	c	44.	d		
22.	b	45.	d		
23.	d	46.	c		

5 SAFETY PROCEDURES

Improvements in the use of proper safety procedures and equipment and increased safety awareness are necessary in the wastewater collection system profession. Based on the number of fatalities and the lost-time accident data compiled each year, statistics show that wastewater collection system operations are second only to underground mining in lost-time accidents. Unfortunately, the possibility of becoming number one in this regard is increasing.

The intent of this section is to increase your hazard awareness and to identify the hazards that you face as operators every day. It will also identify the correct methods that should be used at all times.

1. An operator should never enter a manhole without

    a) Steel-toed shoes
    b) At least one trained coworker at the point of entry to assist
    c) The proper atmospheric metering equipment
    d) Both b and c

2. For personnel receiving a chemical burn upon contact with chlorine gas, it is best to

    a) Move the victim into fresh air
    b) Remove contaminated clothing
    c) Flush affected areas with large amounts of water
    d) All of the above

3. A sewer manhole would be safe to enter after

    a) Testing the atmosphere for suitable conditions
    b) Setting up proper safety precautions topside
    c) Securing the worker going into the manhole with all the proper safety equipment
    d) All of the above

4. Manhole covers may be safely lifted with

    a) Your back
    b) A pole
    c) A screwdriver
    d) A manhole hook

5. Artificial respiration must be started on a nonbreathing victim within

    a) 6 minutes
    b) 5 minutes
    c) 4 minutes
    d) 3 minutes

6. The normal procedure before manhole entry would be to

    a) Provide for traffic control (if applicable)
    b) Monitor atmosphere
    c) Secure workers with proper protective gear
    d) All of the above

7. The three basic classes of hazardous atmospheres are

    a) Poisonous, gaseous, toxic
    b) Toxic, hydrogen sulfide, poisonous
    c) Toxic, explosive, oxygen deficient
    d) Gaseous, liquid, combustible

8. When entering a manhole, when should you turn off the atmospheric metering equipment?

    a) After the atmosphere has been tested, it can be shut off
    b) There is no need for it to be left on after 5 minutes
    c) It should not be turned off until the job has been completed; just because gases may not be present at first, it does not mean they will not appear later
    d) It would be a waste of batteries if left on after you initially test the atmosphere

9. Hydrogen sulfide gas is

    a) A gas that has very little odor
    b) A toxic gas that may be found in sewer lines
    c) A gas that is used for fuel in sewer line construction
    d) All of the above

10. The following is a list of chemical compounds sometimes found in the collection system. Match the correct name to the chemical symbol by placing the correct letter in the space provided.

   a) Carbon dioxide CH_4
   b) Hydrogen sulfide CO_2
   c) Carbon monoxide CO
   d) Methane H_2S

11. Of the following, which statement about sewer gas is *not* correct?

   a) It often accumulates in concentrations sufficient to kill those exposed
   b) When mixed with air, it forms a dangerous explosive mixture
   c) It is composed mainly of hydrogen sulfide and oxygen
   d) None of the above

12. To minimize the danger of coming in contact with infectious diseases, you should

   a) Wear your work clothes home
   b) Always wash your hands after any potential contact with wastewater
   c) Never wear protective clothing when you will come in contact with wastewater
   d) Never repair equipment that comes in contact with wastewater solids

13. When the percentage of oxygen in air is less than _____, it can be fatal.

   a) 20%
   b) 25%
   c) 15%
   d) 30%

14. Any one of the following sewer gases may be present in a manhole or lift station. Which one is most toxic to humans?

   a) Methane
   b) Nitrogen
   c) Carbon dioxide
   d) Hydrogen sulfide

15. Gasoline or volatile organic solvents are objectionable when present in a sewer because

   a)   They present a serious explosion hazard

   b)   They tend to cause corrosion

   c)   They tend to precipitate the wastewater solids, which will cause a stoppage

   d)   They increase the resistance to flow and, therefore, decrease the effective capacity of the system

16. The purpose of a safety meeting is to

   a)   Get off work

   b)   Discuss vacation plans

   c)   Determine how to maximize the life of equipment and tools

   d)   Review potential safety hazards and outline the necessary precautions

17. The most frequent cause of fatal accidents in collection systems is

   a)   Drowning

   b)   Explosions

   c)   Asphyxiation

   d)   Acid inhalation

18. What is the safest method of storing oily rags?

   a)   Hang them up on a nail

   b)   Store them in a wooden barrel

   c)   Keep them in a covered metal container

   d)   Leave them folded in a neat pile on the workbench

19. An accumulation of methane gas in a manhole may create which of the following?

   a)   Asphyxiation hazards

   b)   An explosive mixture

   c)   An oxygen-deficient atmosphere

   d)   All of the above

20. A contributing cause of practically every accident is

   a)   Improper use of tools

   b)   The lack of unity among workers

   c)   Failure to give close attention to the job at hand

   d)   None of the above

21. When excavating a street, signs and barricades are needed

    a)   For the duration of any excavation
    b)   Only when directed by the engineer
    c)   Only if half of the street is to be cut
    d)   Only if the opening will be left overnight

22. When working around backhoes and other moving equipment, the best safety rule to follow is

    a)   Keep your eyes on the equipment at all times
    b)   Stay behind the equipment at all times
    c)   Stay in front of the equipment at all times
    d)   Stay out of the machinery's reach

23. For the protection of operators working in deep sewers, the minimum number of operators who should be stationed on the ground level as a safety measure is

    a)   One
    b)   Two
    c)   Three
    d)   Four

24. In a 12 ft (3.7 m) deep manhole, any person entering the manhole should wear a safety harness and lifeline

    a)   At all times
    b)   When forced draft ventilation is not available
    c)   When less than 18% oxygen is present
    d)   When an explosive gas is detected

25. A good first aid treatment for a burn is to

    a)   Pour cold water on it
    b)   Wrap the area to keep it warm
    c)   Leave the burn alone until you see a doctor
    d)   Apply oil to the burn until you can see a doctor

26. The success of a good safety program depends on the enthusiasm and discipline shown by

    a) The public
    b) The foreman
    c) The supervisors
    d) The individual workers

27. In most jurisdictions, strobe lights are permitted on public utility vehicles when

    a) At the scene of an accident
    b) Towing portable equipment
    c) Traveling to an emergency
    d) All of the above

28. The sense of smell should not be used as a safety tool because

    a) Hydrogen sulfide gas is odorless
    b) One odor may mask the odor of poisonous gas
    c) Some poisonous gases are hard to identify
    d) Both b and c

29. The specific gravity of hydrogen sulfide is 1.19 (air = 1.0). At what place in a manhole would you be most likely to encounter this gas?

    a) Near the top
    b) Closer to the bottom
    c) It would be well mixed with the other gases present
    d) None of the above

30. Puncture wounds are subject to special dangers and are much more likely to become infected than an open wound because

    a) The patient will be suffering from shock
    b) There are generally fractures near the wound
    c) It is hard to stop the flow of blood from a puncture wound
    d) Puncture wounds typically do not bleed freely, so the natural cleansing given by the bleeding does not occur

31. A gas detector or indicator should be used

    a) When entering every other manhole

    b) Before entering all manholes

    c) During the summer only

    d) Only by the supervisor

32. If a worker is overcome by sewer gas, the best first aid procedure while waiting for medical help to arrive is

    a) Cover the person and wait for the doctor

    b) Sponge his/her face with cold water

    c) Leave him/her where they are and apply artificial respiration

    d) Move him/her into fresh air and give him/her artificial respiration if needed

33. Which of the following statements is not true?

    a) At least two people must be available at the ground surface to remove a person wearing a safety harness out of a manhole

    b) A safety harness should be worn by all persons who enter a manhole

    c) If the person in the manhole loses consciousness, one person should enter the manhole to give first aid while the second person goes for help

    d) If the person in the manhole loses consciousness, he/she should be pulled from the manhole immediately

34. Under normal conditions, when equipment is used as a resuscitator, what "gas" is forced into the lungs?

    a) Oxygen only

    b) Nitrogen only

    c) Carbon dioxide

    d) 50% oxygen and 50% carbon monoxide

35. Manhole steps should

    a) Never be used; a ladder is always preferred

    b) Be a minimum of 0.25-in. thick (or approximately 6.5 mm) to prevent corrosion damage weakening them

    c) Never be trusted and carefully inspected before use or a ladder should be used when entering a manhole

    d) Always be used when entering a manhole to reduce chances of an accident

36. How should drinking water be dispensed on a utility truck?

   a) By using a dipper
   b) With glasses
   c) By using a sterile faucet
   d) By paper cups

37. If the forced draft ventilation fan in a pumping station fails, it should be fixed immediately because

   a) Gases could accumulate
   b) The pumps will overheat
   c) There will be less air available for personnel
   d) Both a and c

38. The following gases may be found in sewers:

   a) Flammable explosive gases
   b) Hydrogen sulfate
   c) Carbon tetrachloride
   d) All of the above

39. An oxygen deficiency or dangerous concentration of toxic or suffocating gases could be present in

   a) Manholes that are well ventilated
   b) Wet wells that are well ventilated
   c) Pump rooms that are well ventilated
   d) All of the above

40. Of the following, the most important reason for investigating the cause of an accident is to

   a) Prevent occurrence of the accident again
   b) Determine if the employee injured was at fault
   c) Impress upon the employees the need to be safety minded
   d) Determine if the injured employee deserves compensation

41. Chlorine cylinders (150 lb or 68 kg)

    a) Can be handled safely

    b) Should be stored at temperatures above 50 °F (10 °C) and kept away from steam pipes

    c) Contain a fusible metal safety plug

    d) All of the above

42. The most effective means of reducing atmospheric hazards in a manhole is through the use of

    a) Enzymes

    b) Explosimeters

    c) Masking agents

    d) Portable blowers

43. Chlorinators should be located

    a) In a cool room, typically between 40 to 55 °F (approximately 5 to 13 °C)

    b) In a room where the operators work so they may keep a close watch on the equipment

    c) In a separate room

    d) None of the above

44. When venting a manhole before entry, care should be taken to be sure that

    a) Only cold air is blown into the manhole

    b) Only warm air is blown into the manhole

    c) Truck or blower exhaust gases do not get drawn into the blower intake

    d) Air is not blowing into the basements of homes

45. Which of the following hazards can potentially be encountered when entering a manhole?

    a) Atmospheric hazards

    b) Drowning hazards

    c) Waterborne diseases

    d) All of the above

46. A _____ gasket should always be used on liquid chlorine lines to prevent leaks of this dangerous gas.

    a) Lead
    b) Rubber
    c) Teflon
    d) Any of the above

47. It is safe to enter a manhole after

    a) Testing for gas
    b) Testing for explosive conditions
    c) Testing for toxic, explosive, or oxygen-deficient conditions
    d) Sending a trained worker ahead into the manhole to make sure it is safe

48. Which of the following is responsible for the most lost-time accidents in the wastewater collection field?

    a) Sewer gas explosions
    b) Inhalation of toxic gases
    c) Sewer trench cave-ins
    d) Back injuries caused lifting heavy objects improperly

49. At what location in a room does a chlorine leak detector withdraw a sample of air?

    a) At the ceiling
    b) 3 ft (0.9 m) from the bottom of the floor
    c) At floor level
    d) None of the above

50. Which of the following is least important before entering a manhole or wet well at a lift station?

    a) Provide a forced draft of fresh air into the confined space
    b) Check for the possible presence of waterborne-disease-producing bacteria
    c) Wear a tested safety harness with safety ropes extending to the outside of the manhole
    d) Be sure that appropriate help is available at the surface to assist in an emergency

51. Of the following diseases listed below, which one may be spread by the improper handling of wastewater?

a) Anthrax

b) Malaria

c) Typhoid

d) Sinusitis

52. Under ordinary atmospheric conditions, chlorine gas is

a) Bluish-white

b) Greenish-yellow

c) White

d) Colorless

53. Which of the following should be performed before entering a manhole?

a) Test for the presence of nitrogen gas

b) Check for explosive gases

c) Provide a safety belt for use by at least two people remaining above ground

d) All of the above

54. Assume an accident has occurred, leaving you with an injured worker. The person has passed out and is turning blue. What should you do first?

a) Cover the person with a blanket

b) Check the person's heartbeat, breathing, and begin CPR if necessary

c) Revive with spirits

d) Use ammonia capsules to wake the victim up so he/she can begin to breathe

55. If a highly toxic substance is discovered in the wastewater collection system, what should be done first?

a) Determine the source immediately

b) Take a sample to the laboratory for testing

c) Notify the water resource recovery facility downstream

d) Evacuate the surrounding area and notify the proper authorities

56. Chemicals should not be combined indiscriminately because

    a)   They may be rendered useless
    b)   Explosions or other chemical reactions could result
    c)   The operator may get skin rashes
    d)   Precipitates could form and block the sewer lines

57. Chlorine gas leaks can be easily detected by

    a)   Applying a water-soaked rag over the chlorine equipment
    b)   Passing an open bottle of chloride trioxide along the chlorine equipment
    c)   Passing an open bottle of ammonia along the underside of chlorine equipment
    d)   Using the sniff test

58. Chlorine may be used to

    a)   Control corrosion
    b)   Control odor
    c)   Disinfect treated wastewater
    d)   All of the above

59. Hydrogen sulfide gas smells like which of the following?

    a)   A dead dog
    b)   Gasoline
    c)   Fresh paint
    d)   Hard-boiled eggs

60. Match the following accident hazards with the appropriate safety device/procedure:

_____ 1. Infection	a. Ion-exchange unit
_____ 2. Toxic dust	b. Separate storage rooms
_____ 3. Asphyxiation	c. Methane gas meter
_____ 4. Falls	d. Oil separator
_____ 5. Combustible chemicals	e. Filter respirator
_____ 6. Electric shock	f. Third wire
_____ 7. Explosions	g. Safety belts
	h. Hand washing
	i. Oxygen-deficiency meter

61. Carbon dioxide, one of the gases given off by anaerobic conditions in sewers, is not poisonous, but it may cause

    a) Dizziness

    b) Erratic breathing

    c) Asphyxiation

    d) All of the above

62. Oxygen-deficient atmospheres are a result of the oxygen either being used up or being replaced by another gas. Oxygen can be used up in a confined space by

    a) Fire or combustion

    b) Rusting or scaling of metals

    c) Aerobic bacterial action

    d) Both a and c

63. When an emergency situation occurs, the proper first step is to

    a) Call for help on the radio

    b) Put on protective gear so that you are ready to provide assistance

    c) Clear the area of unnecessary equipment that might interfere with rescue

    d) Get an emergency first aid kit from the truck

64. A worker notices that a necessary piece of equipment is defective. The current job needs to be finished as soon as possible. The worker should

    a) Consult with the other employees and reach a consensus agreement on what to do

    b) Continue the job, but be cautious while using the defective equipment

    c) Report the defect and hold off completing the job until the equipment is repaired or replaced

    d) Stop all work and wait for the boss to come by the job site and make a decision

65. Which of the following are symptoms of oxygen deficiency?

    a) Loss of fine motor control, confusion, difficulty breathing, and ringing in the ears

    b) Extreme thirst, difficulty breathing, pain in arms and ears, and nosebleeds

    c) Shakes, chills, numbing of the extremities, and nosebleeds

    d) None of the above

66. Before setting up for entry into a confined space, equipment must be checked to establish that everything is in working order. When calibrating a gas-monitoring device with fresh air, the lower explosive limit (LEL) display should read

    a) 0% LEL
    b) 10% LEL
    c) 50% LEL
    d) 100% LEL

67. A scope of work for which a confined-space entry permit is authorized can only be changed

    a) When the crew can accomplish additional tasks during the entry at no added cost to the department
    b) When authorized by the permit issuer
    c) When the crew is fully competent to do the added tasks
    d) When authorized by the entry supervisor

68. Notable features of a confined space is/are

    a) Any enclosed area in the work place not designed for continuous human occupancy
    b) Large enough to enter and work in
    c) Has limited or restricted openings that can make the space difficult to enter or exit
    d) All of the above

69. Which of the following is not a characteristic of a permit-required confined space?

    a) Has, or has the potential for, a hazardous atmosphere
    b) Contains, or may contain, material that can surround or engulf you
    c) Has an internal configuration that could cause you to be trapped or asphyxiated
    d) Is only accessible by a vertical entry

70. An LEL reading of 15% on a gas meter means

    a) There is a 15% chance of explosion
    b) The top 15% of the atmosphere is explosive
    c) The atmosphere is safe, no ventilation is needed
    d) The atmosphere is unsafe, ventilate before entering

71. You are at a permit-required confined space. You have broken down your tripod and notice that tools were left in the hole. Your supervisor, who just arrived on site, says to skip the tripod and entry procedures, climb in quickly, and get the tools and get out. You should

  a) Do what the supervisor says; he/she is the boss
  b) Skip the procedures if you want; it is your life
  c) Insist on following the proper procedures or leave the tools
  d) Ask the supervisor to send someone else

Answers for Section 5—Safety Procedures

1.	d	24.	a	47.	c	70.	d
2.	d	25.	a	48.	d	71.	c
3.	d	26.	c	49.	c		
4.	d	27.	d	50.	b		
5.	c	28.	d	51.	c		
6.	d	29.	b	52.	b		
7.	c	30.	d	53.	b		
8.	c	31.	b	54.	b		
9.	b	32.	d	55.	c		
10.	*	33.	c	56.	b		
11.	d	34.	a	57.	c		
12.	b	35.	c	58.	d		
13.	c	36.	d	59.	d		
14.	d	37.	d	60.	**		
15.	a	38.	a	61.	d		
16.	d	39.	d	62.	d		
17.	c	40.	a	63.	a		
18.	c	41.	d	64.	c		
19.	d	42.	a	65.	a		
20.	c	43.	c	66.	a		
21.	a	44.	c	67.	b		
22.	d	45.	d	68.	d		
23.	b	46.	a	69.	d		

*a. Carbon dioxide CO_2
 b. Hydrogen sulfide H_2S
 c. Carbon monoxide CO
 d. Methane CH_4

**1. Infection—h. Hand washing
  2. Toxic dust—e. Filter respirator
  3. Asphyxiation—i. Oxygen deficiency meter
  4. Falls—g. Safety belts
  5. Combustible chemicals—b. Separate storage rooms
  6. Electric shock—f. Third wire
  7. Explosions—c. Methane gas meter

6 DESIGN AND NEW CONSTRUCTION

Many communities design their smaller-diameter sewers in-house and may perform all of their own sewer construction. This section covers some of the pertinent design features, with an emphasis on safety precautions during construction. Some design and construction details may be regional in nature; however, safety features and practices should be standard.

1. Shoring and trenching are most likely to be required under the following condition(s):

    a) For deep cuts in sandy soils
    b) Where a high groundwater table is present
    c) For shallow trenches in muck
    d) All of the above

2. The ideal pipe bedding material is

    a) Gravel
    b) Wet material
    c) Clay
    d) Sand

3. Collection systems must be inspected during construction to

    a) Keep inspectors busy at all times
    b) Determine exfiltration rates
    c) Determine if the contractor is complying with the plans and specifications
    d) Train new inspectors

4. Sewer pipe joints are typically sealed by using

    a) A-rings
    b) O-rings
    c) P-rings
    d) Z-rings

5. What types of pipe are commonly used for sanitary sewers?

    a) Vitrified clay
    b) Polyvinyl chloride (PVC)
    c) Reinforced concrete
    d) All of the above

6. Angle of repose is best defined as

    a) The length of the trench plus the width divided by height of the spoil
    b) The depth of the trench times one-half the height of the spoil
    c) The maximum natural slope of spoil
    d) None of the above

7. Any trench that is more than _____ deep must be shored.

    a) 5 ft (1.5 m)
    b) 6 ft (1.8 m)
    c) 8 ft (2.4 m)
    d) 10 ft (3.1 m)

8. A sewer that is 89.0 ft (27.1 m) MSL means

    a) The top of the sewer is 89.0 ft (27.1 m) above mean sea level
    b) The bottom of the interior diameter is 89.0 ft (27.1 m) above mean sea level
    c) The center of the pipe is at 89.0 ft (27.1 m) elevation
    d) The surface of the wastewater will flow at 89.0 ft (27.1 m) above mean sea level

9. The process by which water seeps or passes through soil is called

    a) Runoff
    b) Seepage
    c) Percolation
    d) Infiltration

10. In accordance with good engineering practice, sanitary sewer lines should be laid at

    a) The steepest grade possible to prevent deposits and subsequent clogging of the lines
    b) A fixed grade of 0.25 in./ft (20.8 mm/m)
    c) A grade that varies with the quantity of wastewater to be carried and pipe diameter
    d) A grade parallel to the surface topography

11. Which of the following materials is not commonly used in the construction of sewer lines?

    a)   Aluminum
    b)   Vitrified clay
    c)   Cast iron and steel
    d)   Precast and monolithic concrete

12. The velocity of flow in a force main during "pump-on" condition should be approximately

    a)   0 to 1 ft/sec (0 to 0.3 m/s)
    b)   1 to 3 ft/sec (0.3 to 0.9 m/s)
    c)   3 to 6 ft/sec (0.9 to 1.8 m/s)
    d)   9 to 12 ft/sec (2.7 to 3.7 m/s)

13. In repairing or constructing collection system lines, which of the following safety equipment should be on hand?

    a)   Hard hats and safety-toed shoes with nonskid composition-type soles
    b)   Safety harness with safety rope
    c)   Adequate bracing and shoring to prevent cave-ins and/or rock slides
    d)   All of the above

14. Clear water from foundation drains, cistern overflows, roof drains, and similar sources

    a)   May be connected to either storm or sanitary sewers, whichever is most convenient
    b)   Should be connected to sanitary sewers because it dilutes the wastewater and makes it easier to treat
    c)   Should not be connected to sanitary sewers because it tends to hydraulically overload the collection system and water resource recovery facility
    d)   None of the above

15. Newly constructed collection systems are typically inspected to ensure

    a)   That the contractor has complied with the plans and specifications
    b)   There is no infiltration
    c)   The municipality receives a working system from the contractor
    d)   All of the above

16. The minimum design velocity for wastewater in a sanitary sewer is

   a) 1 ft/sec (0.4 m/s)
   b) 2 ft/sec (0.6 m/s)
   c) 3 ft/sec (0.9 m/s)
   d) 4 ft/sec (1.2 m/s)

17. Laying sewer lines on proper grades is essential so the wastewater flows fast enough to prevent solids from settling. When the grades are too flat, the wastewater typically becomes

   a) Odorless
   b) Septic
   c) Less objectionable
   d) More easily treated

18. During the installation of a sewer, which of the following pieces of equipment is not used?

   a) Clamshell
   b) Draglines
   c) Front end loader
   d) Front line shell

19. The design of a sanitary collection system should consider

   a) The largest amount of rainfall in a single storm
   b) The number of people using the system
   c) The size of the drainage area
   d) The frequency of the storms

20. Sewers are designed to provide a velocity of

   a) 2 to 10 gpm (7.6 to 37.9 L/min)
   b) 5 to 15 ft/sec (1.5 to 4.6 m/s)
   c) 2 to 10 ft/sec (0.6 to 3.0 m/s)
   d) 15 to 26 ft/sec (4.6 to 7.6 m/s)

21. Material excavated from trenches should be placed no less than

    a) 2 ft (0.6 m) away from the edge of the trench
    b) 3 ft (0.9 m) away from the edge of the trench
    c) 4 ft (1.2 m) away from the edge of the trench
    d) 5 ft (1.5 m) away from the edge of the trench

22. Slope of a sewer may be defined as

    a) The rise over the run
    b) The hydraulic grade divided by slope
    c) The rim elevation minus the invert elevation
    d) None of the above

23. Hydraulic gradient means

    a) The grade of the ground surface
    b) The gradient of the pressure heads
    c) The inclination of the sewer crown
    d) The free surface slope of a liquid in a pipeline

24. When the trench bottom uncovered at a subgrade is so soft that it could not safely support the pipe, it would typically be best to

    a) Use wooden supports for pipe while laying
    b) Fill with the approved material and tamp
    c) Excavate further and refill with approved material
    d) Firmly tamp the material until the required degree of compaction has been reached

25. Which of the following would be best used to control large quantities of water when excavating sewer trenches in sand?

    a) A sump pump
    b) Well points
    c) Cofferdams
    d) Sheet piling

26. When reviewing the design of a pumping station wet well, you should consider

    a)   The slope of the floor
    b)   The height of the walls
    c)   The temperature of the wastewater
    d)   The amount of solids in the wastewater

27. The total dynamic head of flowing liquid is

    a)   The velocity head plus the static head
    b)   The static head plus the miscellaneous head
    c)   The velocity head plus the static head plus the miscellaneous head during pumping
    d)   None of the above

28. The essential requirement of most provisions and specifications for backfilling sewer and storm drain trenches is to

    a)   Reduce the overburden on the conduits
    b)   Prevent displacement damage to the newly laid pipe
    c)   Load the line up to the design loading to see if it will fail
    d)   Load the line to a specified value larger than the loads expected

29. What is the change in elevation between the upstream invert and the downstream invert on a 400-ft (122-m) section of 8-in. (200-mm) sewer pipe with a 4.5% slope?

    a)   18 ft (5.5 m)
    b)   19 ft (5.8 m)
    c)   20 ft (6.1 m)
    d)   Not enough information given

30. When excavating a trench in an existing roadway, it is a good idea to segregate the asphalt from the rest of the material removed from the trench for the following reason(s):

    a)   Because asphalt can be reused after it is melted down
    b)   Because asphalt can be used as ballast during the winter months
    c)   So asphalt is not used for backfill
    d)   Both a and b

31. Barricades and lanterns should

    a)   Never be used on sidewalks

    b)   Always be placed 10 ft (3.1 m) from any excavation

    c)   Be used to keep citizens from inspecting the work

    d)   Be far enough from the excavation to be seen and to allow fast-moving traffic time to slow down

32. Material excavated from a trench should always be

    a)   Removed from the site

    b)   Piled back at least 2 ft (0.6 m) from the trench

    c)   Piled close along the sides of the trench

    d)   Piled on the street side of the trench whenever possible

33. The term applied to bracing of a trench wall during excavation is

    a)   Heading

    b)   Trenching

    c)   Well pointing

    d)   Shoring or sheeting

34. A keyway is used in concrete construction to

    a)   Check the line and grade

    b)   Make an entry into an old brick sewer

    c)   Divert water from a sewer that is being repaired

    d)   Bond freshly poured concrete to concrete that has already set up

35. A grade that has a rise of 6 ft in 600 ft (1.8 m in 183 m) has a _____ grade that is

    a)   0.1%

    b)   1%

    c)   2%

    d)   10%

36. An eccentric manhole cone is

    a)   An extra-wide manhole cone

    b)   A cone that tapers non-uniformly from barrel to manhole cover, with one side typically vertical

    c)   A cone that is used to warn oncoming traffic that there is a manhole open

    d)   There is no such thing as an eccentric manhole

37. Caulking means

   a) Roughing in measurements
   b) Making a joint tight using plaster
   c) Joining two pieces of pipe, one of which has a bell or spigot, by means of packed oakum or lead
   d) All of the above

38. In an older collection system, sewer pipes that are 36 in. (910 mm) and smaller will commonly be made of

   a) Cast iron
   b) Vitrified clay
   c) Asbestos cement
   d) Both c and d

39. The distance of 10.75 ft (3.28 m) is the same as

   a) 10 ft 9 in. (3 m 28 cm)
   b) 10 ft 7½ in. (3 m ³¹/₈ cm)
   c) 10 ft ¾ in. (3 m ¾ cm)
   d) None of the above

40. The reason for digging bell holes for pipe is to

   a) Allow the water to run through
   b) Make the pipe level
   c) Give the pipe even bearing between the joints
   d) All of the above

41. You are reviewing the plans for a new pumping station to be built in your town. The plans call for a spiral staircase as the means of access to the floor of the dry well. One safety objection to this setup might be

   a) The treads on spiral stairs are always more slippery than on conventional rectangular steps
   b) Spiral stairs are typically not structurally sound
   c) It is almost impossible to carry any large individual up or down a spiral staircase in a station
   d) A spiral staircase cannot be built with intermediate landings to provide safe access to intermediate pump shaft bearings

42. A 12-in. (300-mm) diameter pipe is being laid in a trench. How wide should this trench be?

   a)   1.5 ft (0.5 m)
   b)   2.5 ft (0.8 m)
   c)   3.5 ft (1.1 m)
   d)   As narrow as possible as long as it does not interfere with the laying of pipe

43. The method of laying pipe has much to do with the strength developed by the pipe. Which of the four methods listed below has the highest load factor?

   a)   Earth embedment
   b)   Earth embedment with a shaped bottom
   c)   Crushed stone embedment
   d)   Crushed stone embedment with a shaped bottom

44. The gradient or slope of small sewers should be _____ the slope for a large sewer.

   a)   The same as
   b)   Less than
   c)   Greater than
   d)   It does not matter

45. Lines between manholes should have only one grade and be straight because

   a)   Velocity will be maintained
   b)   It is cheaper during construction
   c)   Cleaning will be easier
   d)   All of the above

46. While backfilling a sewer trench with soil, you should

   a)   Do it the easiest way possible
   b)   Backfill completely then tamp down the topsoil
   c)   Backfill in 4-in. (approximately 100-mm) layers and tamp down until the pipe is covered with 12 in. (approximately 300 mm) of backfill
   d)   You must use specification gravel; never use the old soil

47. Which of these factors is not considered when obtaining an easement for construction and maintenance of a collection system?

   a)   Compaction
   b)   Deposition of cut materials
   c)   Existing landscaping
   d)   Vehicle access

Answers for Section 6—Design and New Construction

#	Ans	#	Ans	#	Ans
1.	d	19.	b	37.	c
2.	a	20.	c	38.	d
3.	c	21.	a	39.	a
4.	b	22.	a	40.	c
5.	d	23.	d	41.	c
6.	c	24.	c	42.	d
7.	a	25.	b	43.	d
8.	b	26.	a	44.	c
9.	c	27.	c	45.	d
10.	c	28.	b	46.	c
11.	a	29.	a	47.	b
12.	c	30.	c		
13.	d	31.	d		
14.	c	32.	b		
15.	d	33.	d		
16.	b	34.	d		
17.	b	35.	b		
18.	d	36.	b		

7 ELECTRICAL PUMPS AND MOTORS

Personnel responsible for the management and operation and maintenance of wastewater collection systems must have a basic knowledge of electricity, including several types of pumps and electric motors. The most common type of pumps used in pumping/lift stations are centrifugal, submersible, and, in some cases, airlift for smaller flows. The following questions are to be used as guides only. In no way are wastewater collection system operators encouraged to attempt electrical maintenance. That work should only be performed by qualified electricians.

1. The purpose of a main circuit breaker is to

   a) Provide a complete power disconnect
   b) Ensure there are no power surges
   c) Protect a motor from overheating
   d) Protect the electrical windings of a pump

2. Of the situations and factors listed below, which would lead to the most wear on pump impellers?

   a) Velocity head
   b) Feet of head
   c) Running the pump dry
   d) Grit in wastewater

3. Electrical power would not be required for which of the following pieces of equipment?

   a) An exhaust fan
   b) Pump alternator
   c) A jet flushing system
   d) A light system

4. You have just reinstalled a three-phase, 220-V electric motor with new windings. It starts to run in the wrong direction. The corrective action would be to

    a) Check the motor starter
    b) Reground the ground wire
    c) Reverse the connection of any two motor leads
    d) Send it back to the repair shop to have the windings properly installed

5. Calling in professionals to do electrical work is

    a) Admitting you are not qualified to perform the work
    b) Wasteful when working with budgeted funds
    c) The intelligent choice when considering electrical maintenance duties
    d) None of the above

6. Several types of pumps are used in wastewater collection and treatment. A reciprocating pump

    a) Has a rotating impeller
    b) Has a piston that moves back and forth
    c) Has two valves
    d) Is used to pump grit

7. Which of the following should be observed on newly installed pumps?

    a) Capacity
    b) Vibration
    c) Direction of rotation
    d) All of the above

8. An instrument used to measure resistance in an electrical circuit is a/an

    a) Ammeter
    b) Ohm meter
    c) Mega meter
    d) Voltage tester

9. A lantern ring is

    a) A clasp used to hang a flashlight on when descending a manhole
    b) A device used in the seal housing of a pump
    c) An emergency drop light that casts a brilliant white ring of light
    d) None of the above

10. Wearing rings are installed

    a)   To protect against cavitation

    b)   To wear rather than the impeller

    c)   To protect the bearings from large solids

    d)   Both a and c

11. Grit facilities are typically located upstream of influent pumps to

    a)   Prevent wear on the impellers

    b)   Prevent wear on the pump volutes

    c)   Prevent wear on the pump motor

    d)   Both a and b

12. Which of the following conditions could cause the electrical motor of a pump not to start?

    a)   No power supply

    b)   Fuse or circuit breaker is out

    c)   Impeller turns in the wrong direction

    d)   Both a and b

13. The term *kilowatt-hours* is best applied to

    a)   Power

    b)   Energy

    c)   Capacity

    d)   Resistance

14. Of the following, who would be best qualified to perform the work in an electrical panel if problems are encountered?

    a)   Crew leader

    b)   Electrician

    c)   Supervisor

    d)   Maintenance personnel

15. The maximum size of a fuse placed in the electrical circuit with a motor having a low starting load should have a current rating load of

   a) The same as that of the motor
   b) One and one-half times that of the motor
   c) Three times that of the motor
   d) Three-fourths the size of the motor

16. A pump in a collection system is generally used to

   a) Break up solids
   b) Increase the velocity of flow
   c) Lift wastewater to a higher elevation
   d) Force wastewater through a small pipe

17. The most essential aspect of a good preventive maintenance program is

   a) Good equipment
   b) Adequate records
   c) Good supervision
   d) Good maintenance personnel

18. Why should variable speed pumps be used to pump the flow into the headworks of a large water resource recovery facility?

   a) They can pass solids easier
   b) They run smoother and quieter
   c) They handle variable flowrates more efficiently
   d) They maintain constant flow through the facility

19. A slight water leak past the packing around a pump shaft

   a) Should be eliminated
   b) Means the shaft is worn
   c) Cannot be eliminated
   d) Is essential for proper operation

20. Voltage is essentially

   a) The amount of current drawn
   b) Electrical pressure available
   c) A measure of electrical consumption
   d) None of the above

21. A volute is

    a) A pump assembly
    b) A pump casing
    c) A motor drive
    d) A motor casing

22. Electrical motors draw more power starting than during normal operation conditions because

    a) The overload switch must be deactivated
    b) The motor and the "load" attached to it must start moving
    c) Time is required for the voltage to activate the magnetic field
    d) This statement is incorrect

23. If a fuse continues to blow, an appropriate solution would be to

    a) Provide a "jumper" in the box
    b) Replace it with a higher capacity fuse
    c) Inspect the affected equipment to determine the cause
    d) Replace it with a fuse of lower capacity

24. Why would you shave (turn down) an impeller?

    a) Save energy
    b) Decrease cavitation
    c) Improve efficiency
    d) None of the above

25. Maintenance of couplings between the driving and driven elements includes

    a) Keeping proper curvature
    b) Keeping the electrodes greased
    c) Draining the old oil in fast couplings
    d) Keeping proper alignment, even with flexible couplings

26. Mechanical seals have

    a) Two rotating sections
    b) Two stationary sections
    c) One rotating and one stationary section
    d) None of the above

27. There are two types of shaft alignment: _____ and _____

    a) Thrust and parallel
    b) Rotating and angular
    c) Parallel and angular
    d) Tangential and curved

28. *Cavitation* is best defined as

    a) The pitting of a V-belt drive
    b) An electrical term used for resistance
    c) Separations on the iron core inside the coil of a magnetic starter
    d) The formation and collapse of a gas pocket or bubble on the blade of an impeller

29. When a pump motor draws a higher current than its rating, the reasons may include

    a) Motor may be too small
    b) Brushes may be worn
    c) Something may be caught in the impeller
    d) All of the above

30. Which of the following type of pump works on the principle of a decrease in the over-all specific weight of a confined column of a gas-water mixture?

    a) Gear
    b) Piston
    c) Air lift
    d) Centrifugal

31. Which of the following may lead to bearing temperatures exceeding 160 °F (71 °C) on a centrifugal pump?

    a) Overlubrication
    b) Underlubrication
    c) Start of a bearing failure
    d) All of the above

32. A leak around a double mechanical seal of a pump

    a)   Is needed to help cool the seal

    b)   Means the seal must be tightened

    c)   Is an indication of seal failure

    d)   All of the above

33. Excessive lubricant pressure or excessive grease in bearings may be prevented by using

    a)   Drip pans

    b)   Wire brushes

    c)   Relief plugs

    d)   Hydraulic valves

34. If a pump and force main are designed to deliver 300 gpm (18.9 L/s), what will happen if a duplicate pump is added to the system?

    a)   You will pump 600 gpm (37.9 L/s)

    b)   You will pump about 450 gpm (28.4 L/s)

    c)   The amount of flow will depend on the static head loss

    d)   The amount of flow will depend on the static and dynamic head loss

35. The spiral-shaped casing surrounding a pump impeller that collects the water discharged by the impeller is called a

    a)   Sleeve

    b)   Volute

    c)   Cavity

    d)   Flapper valve

36. Electric power may be measured in

    a)   Kilowatts

    b)   Watts

    c)   Amps

    d)   Both a and b

37. If a pump's packing is not properly maintained, the following problems can arise:

  a) Shaft and sleeve damage

  b) Contaminated bearings

  c) Loss of suction because of air being allowed to the pump

  d) Both a and b

38. In operating a small wastewater pumping station with two identical pumps, it is best to adjust the controls so that

  a) The pumps turn on together

  b) The pumps alternate in operation

  c) One pump operates continually and the other pump is held in reserve until the first one is worn out

  d) The pumps operate only when the wastewater has reached a level at which it would overflow the bypass

39. An external source of clean water is used in a packing gland primarily to

  a) Keep the packing from drying out

  b) Lubricate the shaft

  c) Keep gritty material from wearing the shaft sleeve

  d) Cool the packing gland

40. The primary purpose of wastewater lift stations in collections systems is to

  a) Prevent sewers from being laid too deep

  b) Increase the capacity of the sewer line

  c) Increase velocity of wastewater in the sewer line

  d) Lift the wastewater to a higher elevation where it can flow by gravity

41. The electrical load placed on a single circuit

  a) Is not affected by the addition of one more motor

  b) May be increased without limit and no damage will be done

  c) Is of no interest to collection system personnel

  d) Should be limited by the maximum current rating of the wiring and switches

42. The effect of partially closing the discharge valve on a motorized centrifugal pump driven by a three-phase induction motor would be to cause the

    a) Motor to run hotter
    b) Motor to run slower
    c) Pumped wastewater to get colder
    d) None of the above

43. A centrifugal pump in a pumping station is running, but not pumping out the wet well. The pump is hot. What is/are the probable cause(s)?

    a) The pump is air bound
    b) The check valve is stuck in the closed position
    c) The static head in the force main is greater than the discharge head of the pump
    d) All of the above

44. A centrifugal pump operating under _____ can prime itself without any external priming device.

    a) Negative suction head
    b) Positive suction head
    c) A dynamic head
    d) A static head

45. What information must be on a warning tag attached to a switch that has been locked out?

    a) Directions for removing the tag
    b) Name of the supervisor to call in case of an emergency
    c) Time to unlock switch
    d) Signature of person who locked out the switch and who is the only person authorized to remove the tag

Answers for Section 7—Electrical Pumps and Motors

1.	a	24.	c
2.	d	25.	d
3.	c	26.	c
4.	c	27.	c
5.	c	28.	d
6.	b	29.	d
7.	d	30.	c
8.	b	31.	d
9.	b	32.	c
10.	b	33.	c
11.	d	34.	d
12.	d	35.	b
13.	a	36.	d
14.	b	37.	d
15.	b	38.	b
16.	c	39.	b
17.	b	40.	d
18.	c	41.	d
19.	d	42.	a
20.	b	43.	d
21.	b	44.	b
22.	b	45.	d
23.	c		

8 MATHEMATICS

Wastewater collection system operators should have a good working knowledge and understanding of basic mathematics to properly calculate construction costs, operating costs, and chemical dosages and to order the correct quantities of materials. In addition, basic math skills are necessary to effectively and efficiently manage both operating and capital budgets.

The mathematical problems presented in this section have varying degrees of difficulty, but represent the "real world" situations that wastewater collection system operators may encounter as they accept higher levels of responsibility within this profession.

1. A rectangular wet well has the following dimensions:

   - Depth of liquid is 10 ft (3.0 m)
   - Width of tank is 12 ft (3.7 m)
   - Length of tank is 15 ft (4.6 m)

   Approximately how many gallons (liters) are in this wet well?

   a) 2400 gal (9100 L)
   b) 13 500 gal (51 100 L)
   c) 18 000 gal (68 150 L)
   d) 20 800 gal (78 750 L)

2. An 8-in. (200-mm) sewer carries a flow of 0.5 mgd (1 890 000 L/d) when flowing full. What is the average velocity?

   a) 2.0 ft/sec (0.6 m/s)
   b) 2.2 ft/sec (0.7 m/s)
   c) 4.4 ft/sec (1.3 m/s)
   d) 5.0 ft/sec (1.5 m/s)

3. If the velocity of flow in a pipe flowing full is 2 ft/sec (0.6 m/s) and the diameter of the pipe is 1 ft (300 mm), what is the flowrate?

   a) 1.4 gpm (5.31 L/min)
   b) 1.6 cfs (45.3 L/s)
   c) 2.5 cfs (70.8 L/s)
   d) 2.8 mgd (10 600 m³/d)

4. The volume of water necessary to fill a 100-ft (30-m) -long section of 12-in. (300-mm) pipe is approximately

    a) 75 gal (280 L)
    b) 591 gal (2100 L)
    c) 834 gal (3150 L)
    d) 5900 gal (22 300 L)

5. To rebuild a manhole, it will be necessary to remove the asphalt paving from a 35-ft (11-m) diameter circle in the street. The pavement area required to be removed is

    a) 200 sq ft (19 m²)
    b) 314 sq ft (29 m²)
    c) 962 sq ft (95 m²)
    d) 1200 sq ft (111 m²)

6. Given the temperature of wastewater at 50 °F, what is the corresponding temperature in °C?

    a) 7 °C
    b) 10 °C
    c) 14 °C
    d) 24 °C

7. At a flowrate of 24 gpm (91 L/min), how long would it take to fill a basin that is approximately 4 ft (1.2 m) wide by 6 ft (1.8 m) long and 2 ft (0.6 m) deep?

    a) About 5 minutes
    b) About 10 minutes
    c) About 15 minutes
    d) About 20 minutes

8. A sewer has failed and 127 ft (39 m) of 8-in. (200-mm) pipe must be replaced. How many 5-ft (1.5-m) sections of pipe will be replaced?

    a) 5
    b) 16
    c) 25
    d) 26

9. What is the grade if there is a rise of 3 ft (0.9 m) over 300 ft (90 m)?

    a) 0.01%
    b) 0.10%
    c) 1.0%
    d) 3.0%

10. To control hydrogen sulfide and odors in a 16-in. (400-mm) sewer, the chlorine dose must be 10 mg/L. The flow is 0.37 mgd (1400 m³/d). What is the approximate feed rate in pounds per day (lb/d) or kilograms per day (kg/d)?

    a) 15 lb/d (7 kg/d)
    b) 23 lb/d (10 kg/d)
    c) 31 lb/d (14 kg/d)
    d) 36 lb/d (16 kg/d)

11. The length of 15 ft 6 in. (4500 mm) is the same as

    a) 15.4 ft (4.4 m)
    b) 15.5 ft (4.5 m)
    c) 15.6 ft (4.8 m)
    d) 15.7 ft (4.9 m)

12. A single-piston reciprocating pump has a 6-in. (150-mm) diameter piston with a 6-in. (150-mm) length of stroke. If it makes 16 discharge strokes per minute, what is the approximate pumping rate in gallons per minute (gpm) or liters per minute (L/min)?

    a) 6.3 gpm (24 L/min)
    b) 11.7 gpm (48 L/min)
    c) 28.1 gpm (105 L/min)
    d) 33.3 gpm (126 L/min)

13. If a 3-ft (0.9-m) square channel has 2 ft (0.6 m) of water flowing at 4 ft/sec (or 1.2 m/s), what is the flow?

    a) 4.5 cfs (0.13 m³/s)
    b) 12 cfs (0.35 m³/s)
    c) 24 cfs (0.65 m³/s)
    d) 36 cfs (1.02 m³/s)

14. The gauge pressure that is equivalent to the static head from a 75-ft (23-m) column of water is approximately

   a)   19.1 psig (131.7 kPa)
   b)   26.7 psig (184.1 kPa)
   c)   32.5 psig (225.6 kPa)
   d)   37.5 psig (258.6 kPa)

15. How many cubic yards (cu yd) or cubic meters (m³) of earth will be removed to excavate a trench 4 ft (1.2 m) wide, 7 ft (2.1 m) deep, and 40 ft (12.2 m) long?

   a)   27.0 cu yd (20.6 m³)
   b)   32.5 cu yd (24.8 m³)
   c)   41.5 cu yd (30.7 m³)
   d)   48.0 cu yd (36.7 m³)

16. A new manhole has been installed 320 ft (97.5 m) from an existing manhole. How far would this new manhole be located from the old one on a map with a scale of 1 in. (25 mm) equals 50 ft (15 m)?

   a)   6.4 in. (163 mm)
   b)   7.8 in. (198 mm)
   c)   8.9 in. (226 mm)
   d)   12.1 in. (307 mm)

17. Given a biochemical oxygen demand (BOD) of 120 mg/L and a flowrate of 18 mgd (68 130 m³/d), calculate the approximate BOD loading on the system.

   a)   5000 lb/d (2268 kg/d)
   b)   10 000 lb/d (4536 kg/d)
   c)   18 000 lb/d (8175 kg/d)
   d)   19 000 lb/d (8618 kg/d)

18. A trench that is 2.5 ft (0.8 m) wide, 6.5 ft (2.0 m) deep, and 65 ft (20 m) long is to be filled with select backfill. Approximately how many cubic yards (cu yd) or cubic meters (m³) of backfill are required?

   a)   16 cu yd (12 m³)
   b)   32 cu yd (24 m³)
   c)   39 cu yd (32 m³)
   d)   45 cu yd (34 m³)

19. If a wet well is 10 ft (3 m) by 10 ft (3 m) by 10 ft (3 m) and the pump turns on at 9 ft (2.7 m) and shuts off at 3 ft (0.9 m), how much liquid is removed in each cycle?

    a)   3700 gal (14 000 L)

    b)   4488 gal (16 200 L)

    c)   5000 gal (18 925 L)

    d)   5244 gal (19 850 L)

20. Assuming that all the slopes are the same, what combination of two pipes would more closely equal the flow through one 12-in. (300-mm) pipe?

    a)   Two 8-in. (200-mm) pipes

    b)   One 8-in. (200-mm) pipe and one 10-in. (250-mm) pipe

    c)   One 10-in. (250-mm) pipe and one 6-in. (150-mm) pipe

    d)   One 10-in. (250-mm) pipe and one 4-in. (100-mm) pipe

21. A wet well has an area of 400 sq ft (37 m²) and a drawdown of 3 ft (0.9 m). What is the volume of wastewater in gallons contained within the drawdown volume?

    a)   1250 gal (4730 L)

    b)   8976 gal (33 300 L)

    c)   9154 gal (34 650 L)

    d)   20 153 gal (76 280 L)

22. An 8-in. (200-mm) sewer with a slope of 0.026 has a vertical drop of _____ ft (meters) in 300 ft (91 m).

    a)   6.4 ft (1.9 m)

    b)   7.8 ft (2.4 m)

    c)   9.3 ft (2.8 m)

    d)   25.6 ft (7.8 m)

23. A slug of dye was introduced to a sewer manhole. One minute and 15 seconds later, the dye was observed at a manhole 268 ft (82 m) downstream. The approximate average wastewater velocity between these manholes is

    a)   0.28 ft/sec (0.09 m/s)

    b)   3.57 ft/sec (1.1 m/s)

    c)   28.0 ft/min (8.5 m/min)

    d)   3.57 ft/min (1.1 m/min)

24. Given a velocity of 2 ft/sec (0.6 m/s) in a 15-in. (380-mm) sewer line with the pipe flowing half full, the approximate flow in gallons per minute (gpm) or (L/min) is

   a)   200 gpm (757 L/min)

   b)   400 gpm (1514 L/min)

   c)   550 gpm (2160 L/min)

   d)   650 gpm (2460 L/min)

25. What is the capacity of a wet well if the pump is rated at 55 gpm (208 L/min) and requires 1 hour and 40 minutes to empty the wet well?

   a)   2100 gal (7950 L)

   b)   3500 gal (13 250 L)

   c)   4700 gal (17 790 L)

   d)   5500 gal (20 800 L)

26. How efficient is a pump if the output power is 1.77 hp (1.32 kW) and the input (brake) power is 2.50 hp (1.87 kW)?

   a)   58%

   b)   60%

   c)   71%

   d)   75%

27. A sewer trench is 2.5 ft (0.77 m) wide, 1000 ft (306 m) long, and 8 ft (2.4 m) deep.

How many cubic yards (cu yd) or cubic meters (m³) of excavation volume will need to be removed from the trench?

   a)   625 cu yd (478 m³)

   b)   741 cu yd (566 m³)

   c)   789 cu yd (603 m³)

   d)   845 cu yd (646 m³)

28. The invert elevation at manhole A is −203.7 ft (−62.1 m) and 325 ft (99 m) downstream at manhole B the invert elevation is −205.0 ft (−62.5 m). What is the slope or grade of this section of pipe?

   a)   0.003

   b)   0.004

   c)   0.006

   d)   0.008

29. What concentration of chlorine in milligrams per liter is applied to a flow of 3 mgd (11 355 m³/d) if the total weight of 100% chlorine used was 160 lb (72.6 kg)?

    a)   3.8 mg/L
    b)   4.4 mg/L
    c)   6.4 mg/L
    d)   10.7 mg/L

30. A flow of 1150 gpm (4353 L/min) is approximately the same as

    a)   0.5 mgd (1900 m³/d)
    b)   0.9 mgd (3400 m³/d)
    c)   1.2 mgd (4550 m³/d)
    d)   1.7 mgd (6300 m³/d)

31. When compared to the cross-sectional area of an 8-in. (200-mm) pipe, the cross-sectional area of a 6-in. (150-mm) pipe is approximately

    a)   One-quarter the area of the 8-in. (200-mm) pipe
    b)   One-half the area of the 8-in. (200-mm) pipe
    c)   Twice the area of the 8-in. (200-mm) pipe
    d)   Three quarters of the area of the 8-in. (200-mm) pipe

32. The hour meters in a pumping station show a total time for the two pumps running to be 8 hours and 47 minutes during the previous 24-hour period. Calculate the volume of flow through the pumping station for the 24-hour period knowing that the pumps have a capacity of 450 gpm (1703 L/min).

    a)   212 000 gal (802 420 L)
    b)   237 000 gal (897 480 L)
    c)   343 000 gal (1 298 255 L)
    d)   530 000 gal (2 006 050 L)

33. In preparing an estimate for the following job, which is to construct a sanitary sewer 850 ft (259.1 m) long, assume that it will be laid at a depth of 7.5 ft (2.3 m) and the trench will be 2.5 ft (0.8 m) wide. Also assume that it has been determined that the excavation, backfilling, and compaction will cost $15.50 per linear foot ($50.85 per linear meter). The pipe will cost $3.75 per linear foot ($12.30 per linear meter). The asphalt surface that must be replaced will cost $3.20 per linear foot ($10.50 per linear meter), including the base.

A. What is the cost of the pipe for the job?

a) $116
b) $2,720
c) $3,190
d) $13,175

B. What is the cost of the excavating, backfilling, and compaction for the job?

a) $3,200
b) $10,875
c) $13,175
d) $15,250

C. What is the cost of replacing the asphalt surface?

a) $2,720
b) $3,450
c) $7,500
d) $10,240

D. What is the total cost of the job?

a) $7,616
b) $16,342
c) $19,085
d) $28,306

E. What is the total cost per linear foot (meter)?

a) $12.75/ft ($38.50/m)
b) $15.82/ft ($41.82/m)
c) $22.45/ft ($73.66/m)
d) $18.17/ft ($54.51/m)

34. The wet well of a pumping station is 5 ft (1.5 m) wide by 6 ft (1.8 m) long. With only one pump running and discharging 280 gpm (1060 L/min), the wet well level was observed to rise 1 ft (0.3 m) in 3 minutes 15 seconds. What was the rate of flow into the wet well?

    a)   69 gpm (261 L/min)

    b)   147 gpm (556 L/min)

    c)   272 gpm (1030 L/min)

    d)   349 gpm (1306 L/min)

35. Manhole A is at station 475.5 ft (145 m) and manhole B is at station 721.7 ft (220 m). If the upstream elevation at A is 143.21 ft (43.65 m) and the sewer is to have a slope of 0.0036, then the downstream elevation of the invert at B should be

    a)   134.35 ft (40.95 m)

    b)   142.32 ft (43.38 m)

    c)   145.00 ft (44.20 m)

    d)   246.20 ft (75.04 m)

36. A rectangular basin is 18 ft (5.5 m) wide by 50 ft (15.2 m) long. It tapers from a depth of 15 ft (4.6 m) at one end to 12 ft (3.7 m) at the other end. The volume of the tank is

    a)   10 800 cu ft (306 m³)

    b)   12 150 cu ft (347 m³)

    c)   14 080 cu ft (400 m³)

    d)   16 110 cu ft (456 m³)

37. A sewer is to be filled with a root control solution containing 250 mg/L of a specified chemical. How much of the chemical would be needed for a 265-ft (81-m) long section of 8-in. (200-mm) diameter sewer?

    a)   0.17 lb (0.08 kg)

    b)   1.00 lb (0.45 kg)

    c)   1.44 lb (0.64 kg)

    d)   5.25 lb (2.34 kg)

38. How many pounds of chlorine are there in 8 lb (3.6 kg) of calcium hypochlorite (HTH) that has 70% available chlorine?

    a)   5.6 lb (2.5 kg)

    b)   6.8 lb (3.1 kg)

    c)   8.0 lb (3.6 kg)

    d)   9.2 lb (4.2 kg)

39. 4 cfs (or 113 L/s) is equal to

    a)   1.12 mgd (4239 m³/d)

    b)   2.59 mgd (9763 m³/d)

    c)   4.34 mgd (16 427 m³/d)

    d)   7.48 mgd (28 312 m³/d)

40. If a certain pumping station pumps at a rate of 1 700 000 gpd (6 434 500 L/d), how many gallons (liters) are pumped in 7 hours?

    a)   495 831 gal (1 876 729 L)

    b)   692 463 gal (2 620 972 L)

    c)   948 117 gal (3 588 623 L)

    d)   1 700 000 gal (6 434 500 L)

41. A pump that has a capacity of pumping 72 000 gpd (272 520 L/d) based on continuous pumping can deliver how much flow per minute?

    a)   42 gpm (159 L/min)

    b)   45 gpm (170 L/min)

    c)   47 gpm (178 L/min)

    d)   50 gpm (189 L/min)

42. A circular wet well that is 20 ft (6 m) in diameter is filled with wastewater to a depth of 11 ft (3.4 m). How many gallons (liters) of water are in the tank?

    a)   11 784 gal (44 602 L)

    b)   25 851 gal (96 200 L)

    c)   52 491 gal (198 678 L)

    d)   103 343 gal (391 153 L)

43. You are constructing a sewer for a subdivision that will contain 8-in. (200-mm) and 10-in. (250-mm) sewer lines. Given the following data:

- Total in-place cost of 8-in. (200-mm) pipe: $62.50/ft ($205.18/m)
- Total in-place cost of 10-in. (250-mm) pipe: $68.00/ft ($224.40/m)
- Pipe cost equals 20% of in-place cost
- Cost of backfilling, compaction, and so on equals 70% of in-place cost
- The amount of pipe to be installed: 1.2 miles (1.93 km) of 8-in. (200-mm) pipe, 0.4 miles (0.64 km) of 10-in. (250-mm) pipe

A. What is the total cost of the project?

a) $212,445
b) $328,948
c) $539,616
d) $672,432

B. What is the total cost of purchasing the 8-in. (200-mm) and 10-in. (250-mm) pipe?

a) $107,923
b) $111,434
c) $123,892
d) $134,768

C. What is the personnel cost?

a) $48,117
b) $53,962
c) $69,234
d) $75,487

D. What are the backfilling and compaction costs?

a) $117,892
b) $296,421
c) $325,462
d) $377,731

44. You work for a public works department for a small city and you manage a small collection system. The system includes 15 miles (24.14 km) of 8-in. (200-mm) pipe, 3 miles (4.83 km) of 12-in. (300-mm) pipe, 2 miles (3.22 km) of 18-in. (450-mm) pipe, and 1.5 miles (2.41 km) of 24-in. (600-mm) pipe.

The decision has been made to clean the entire system every 5 years. You received the following bid to have the lines cleaned by a contractor:

- $7.50/ft ($24.60/m) for the 8-in. (200-mm) pipe
- $7.75/ft (25.40/m) for the 12-in. (300-mm) pipe
- $8.25/ft ($27.10/m) for the 18-in. (450-mm) pipe
- $9.00/ft ($29.50/m) for the 24-in. (600-mm) pipe

Alternatively, you could do all of the cleaning in-house. To do that, the department of public works would need to purchase a new jet machine. You estimate that it would cost $250,000 for capital cost, interest, inflation, supplies, transportation, and maintenance over the next 5 years to operate the machine. In addition, you would need to pay an employee approximately $24.00/hr (includes wage and overhead) to operate the jet machine. The average cleaning rate for the jet machine is 1200 ft (365 m) per day for 6- to 12-in. (150- to 300-mm) pipe and 1000 ft (305 m) per day for 18- to 24-in. (450- to 600-mm) pipe. Assume that the average work day is 8 hours long.

A. How much would it cost to have a contractor clean the entire collection system?

a) $425,000
b) $594,000
c) $875,000
d) $998,000

B. How much would it cost for equipment and personnel for the city to clean its own collection system?

a) $268,800
b) $338,500
c) $563,480
d) $718,850

C. If you recommend that the city cleans its own collection system rather than contracting out for the services, how much annual average savings would the city realize?

a) $98,760/yr
b) $121,240/yr
c) $245,340/yr
d) $459,710/yr

45. Of the following, the closest decimal equivalent of 1/8 is

a) 0.007
b) 0.125
c) 0.555
d) 0.875

46. A full tank truck with a 1500-gal (5675-L) capacity weighs approximately what when filled with water? The truck weighs approximately 5000 lb (2268 kg).

a) 9500 lb (4309 kg)
b) 11 220 lb (5089 kg)
c) 12 500 lb (5670 kg)
d) 17 500 lb (7943 kg)

47. During a closed-circuit television inspection, it was found that 23 out of 90 joints had excessive infiltration. On a percentage basis, how many of the joints were faulty?

    a) 7.8%
    b) 23.0%
    c) 25.6%
    d) 90.8%

48. A 50-hp (or 37.3-kW) motor runs an average of 18 hours per day. The efficiency of the motor is 84% and electricity costs $0.04/kWh. What will the monthly power cost be if there are 30 days in this particular month?

    (Note: 1 hp = 746 W)

    a) $960
    b) $1,120
    c) $1,850
    d) $2,260

49. A 16-in. (406-mm) force main leading to a water resource recovery facility is 3.5 miles (5.636 km) long. If the wastewater is flowing at 2 ft/sec (0.61 m/s), how long will it take (approximately) for the wastewater to reach the facility?

    a) 1.8 hours
    b) 2.6 hours
    c) 4.0 hours
    d) 4.6 hours

50. A chemical costs $1.35 per 50 lb (22.68 kg). What would be the daily cost to treat 8 mgd (30 280 m^3/d) at a dosage of 12 mg/L?

    a) $13.12/d
    b) $21.62/d
    c) $28.34/d
    d) $562.00/d

51. Using a labor cost of $12.50 per hour, what is the cost of a job that takes two people 10 hours to complete?

    a) $125
    b) $250
    c) $500
    d) $1000

52. A trench 2.5 ft (0.76 m) wide, 6.5 ft (1.98 m) deep, and 65 ft (19.8 m) long is to be filled with select backfill. How many cubic yards (cu yd) or cubic meters (m³) of backfill are required?

    a) 39 cu yd (30 m³)

    b) 78 cu yd (60 m³)

    c) 146 cu yd (112 m³)

    d) 390 cu yd (298 m³)

53. A 15-in. (381-mm)-diameter sewer is flowing full and carrying a flow of 2.7 mgd (10 220 m³/d). The average wastewater velocity should be

    a) 3.4 ft/sec (1.1 m/s)

    b) 4.1 ft/sec (1.25 m/s)

    c) 4.8 ft/sec (1.5 m/s)

    d) 5.1 ft/sec (1.55 m/s)

Answers for Section 8—Mathematics (International System of Units [SI])

1. Volume of wet well = Length × Width × Depth
$$= 3.0 \text{ m} \times 3.7 \text{ m} \times 4.6 \text{ m}$$
$$= 51 \text{ m}^3$$

There are 1000 liters in 1 cubic meter.

$$51 \text{ m}^3 \times \frac{1000 \text{ L}}{\text{m}^3} = 51\ 000 \text{ L}$$

Correct answer: b

2. $$\text{Velocity} = \frac{\text{Flow}}{\text{Area}}$$

$$\text{Flow} = 1\ 890\ 000 \frac{\text{L}}{\text{d}} \times \frac{1 \text{ d}}{24 \text{ h}} \times \frac{1 \text{ h}}{60 \text{ min}} \times \frac{1 \text{ min}}{60 \text{ s}}$$

$$= 21.9 \frac{\text{L}}{\text{s}}$$

There are 1000 liters in 1 cubic meter.

$$\text{Flow} = 21.9 \frac{\text{L}}{\text{s}} \times \frac{1 \text{ m}^3}{1000 \text{ L}} = 0.022 \frac{\text{m}^3}{\text{s}}$$

The pipe diameter is 200 mm. There are 1000 millimeters in 1 meter.

$$\text{The pipe diameter} = 200 \text{ mm} \times \frac{1 \text{ m}}{1000 \text{ mm}} = 0.2 \text{ m}$$

$$\text{Area of the pipe} = \pi \left(\frac{\text{Diam}}{2} \right)^2 = \pi \left(\frac{0.2 \text{ m}}{2} \right)^2 = 0.03 \text{ m}^2$$

$$\text{Velocity} = \frac{\text{Flow}}{\text{Area}} = \frac{0.022 \text{ m}^3/\text{s}}{0.03 \text{ m}^2} = 0.73 \text{ m/s rounded to } 0.7 \text{ m/s}$$

Correct answer: b

3. Flow = Area × Velocity

$$\text{Area of the pipe} = \pi\left(\frac{\text{Diam}}{2}\right)^2 = \pi\left(\frac{0.3 \text{ m}}{2}\right)^2 = 0.07 \text{ m}^2$$

$$= 0.07 \text{ m}^2 \times 0.6\frac{\text{m}}{\text{s}}$$

$$= 0.042 \text{ m}^3/\text{s rounded to } 0.04 \text{ m}^3/\text{s}$$

Correct answer: b

4. Volume = Cross-sectional area × Length

There are 1000 millimeters in 1 meter.

$$300 \text{ mm} \times \frac{1 \text{ m}}{1000 \text{ mm}} = 0.30 \text{ m}$$

$$\text{Area of the pipe} = \pi\left(\frac{\text{Diam}}{2}\right)^2 = \pi\left(\frac{0.30 \text{ m}}{2}\right)^2 = 0.07 \text{ m}^2$$

$$\text{Volume} = 0.07 \text{ m}^2 \times 30 \text{ m} = 2.1 \text{ m}^3$$

There are 1000 liters in 1 cubic meter.

$$2.1 \text{ m}^3 \times \frac{1000 \text{ L}}{\text{m}^3} = 2100 \text{ L}$$

Correct answer: b

5. $$\text{Area} = \pi\left(\frac{\text{Diam}}{2}\right)^2$$

$$\text{Area} = \pi\left(\frac{11 \text{ m}}{2}\right)^2 = 95 \text{ m}^2$$

Correct answer: c

6. Celsius temperature = (5/9) (Fahrenheit temperature − 32)

$$= (5/9) (50 - 32)$$

$$= 10 \text{ °C}$$

Correct answer: b

7. $$\text{Time} = \frac{\text{Volume}}{\text{Flow}}$$

Volume = Length × Width × Depth

$$= 1.8\ \text{m} \times 1.2\ \text{m} \times 0.6\ \text{m} = 1.3\ \text{m}^3$$

One cubic meter holds 1000 liters.

$$1.3\ \cancel{\text{m}^3} \times \frac{1000\ \text{L}}{\cancel{\text{m}^3}} = 1300\ \text{L}$$

$$\text{Time} = \frac{1300\ \cancel{\text{L}}}{91\ \cancel{\text{L}}/\text{min}} = 14.3\ \text{min rounded up to 15 min}$$

Correct answer: c

8. $$\frac{39\ \text{m}}{1.5\ \text{m}} = 26 \text{ pieces of 1.5-m lengths of pipe}$$

Correct answer: d

9. $$\text{Grade} = \frac{\text{Rise}}{\text{Run}}$$

$$= \frac{0.9\ \cancel{\text{m}}}{90\ \cancel{\text{m}}} = 0.01 \times 100\% = 1.0\%$$

Correct answer: c

10. Feed rate = Chlorine dose × Flowrate

$$= 10\ \cancel{\text{mg/L}} \times 1400\frac{\cancel{\text{m}^3}}{\text{d}} \times 0.001\frac{\text{kg}/\cancel{\text{m}^3}}{\cancel{\text{mg/L}}}$$

$$= 14\ \text{kg/d}$$

Correct answer: c

11. There are 1000 millimeters in 1 meter.

$$500\ \cancel{\text{mm}} \times \frac{1\ \text{m}}{1000\ \cancel{\text{mm}}} = 0.5\ \text{m}$$

4 m + 0.5 m = 4.5 m

Correct answer: b

12. Discharge per stroke = Area of piston head × Length of stroke

There are 1000 millimeters in 1 meter.

$$150 \; \cancel{mm} \times \frac{1 \; m}{1000 \; \cancel{mm}} = 0.15 \; m$$

$$\text{Area of piston head} = \pi \left(\frac{Diam}{2} \right)^2 = \pi \left(\frac{0.15 \; m}{2} \right)^2 = 0.02 \; m^2$$

Discharge per stroke = 0.02 m² × 0.15 m = 0.003 m³

One cubic meter holds 1000 liters.

$$0.003 \; \cancel{m^3} \times \frac{1000 \; L}{1 \; \cancel{m^3}} = 3 \; L/\text{stroke}$$

$$3 \frac{L}{\cancel{\text{stroke}}} \times 16 \; \frac{\cancel{\text{stroke}}}{min} = 48 \; L \, / \, min$$

Correct answer: b

13. Flow = Area × Velocity

= (0.9 m × 0.6 m) × 1.2 m/s

= 0.65 m³/s

Correct answer: c

14. Each meter of water column is equivalent to 9.807 kPa.

$$23 \; \cancel{m} \times \frac{9.807 \; kPa}{1 \; \cancel{m}} = 225.6 \; kPa$$

Correct answer: c

15. Volume = Length × Width × Depth

= 12.2 m × 1.2 m × 2.1 m

= 30.7 m³

Correct answer: c

16. $\dfrac{15\ m}{97.5\ m}=\dfrac{25\ mm}{x}$

$15\ m\ x=(97.5\ m)(25\ mm)$

$x=\dfrac{97.5\ \cancel{m}}{15\ \cancel{m}}(25\ mm)$

$x=163\ mm$

Correct answer: a

17.   BOD loading = BOD concentration × Flow

$$=120\ \cancel{mg/L}\times 68\ 130\dfrac{m^{3}}{d}\times 0.001\dfrac{kg/m^{3}}{\cancel{mg/L}}$$

$$=8175.6\ kg/d\ \text{rounded down to 8175 kg/d}$$

Correct answer: c

18.   Volume = Length × Width × Depth

$=20\ m\times 0.8\ m\times 2.0\ m$

$=32\ m^{3}$

Correct answer: c

19.   Volume = Length × Width × Depth

Our pump turns on at a 2.7-m depth and shuts off at a 0.9-m depth.

Therefore, the depth of water removed during each cycle is

2.7 m − 0.9 m = 1.8 m

Volume = 3 m × 3 m × 1.8 m

$=16.2\ m^{3}$

There are 1000 liters in 1 cubic meter.

$16.2\ \cancel{m^{3}}\times\dfrac{1000\ L}{1\ \cancel{m^{3}}}=16\ 200\ L$

Correct answer: b

20. The cross-sectional area of a 300-mm diameter pipe is

There are 1000 millimeters in 1 meter.

$$300 \; \cancel{mm} \times \frac{1 \, m}{1000 \; \cancel{mm}} = 0.3 \, m$$

$$Area = \pi \left(\frac{Diam}{2} \right)^2 = \pi \left(\frac{0.3 \, m}{2} \right)^2$$

$$= 0.07 \, m^2$$

The area of which combination of smaller pipes is most closely equal to this?

The area of two 200-mm pipes (Answer a) is

$$Area \; of \; 200\text{-}mm \; pipe = \pi \left(\frac{Diam}{2} \right)^2 = \pi \left(\frac{0.2 \, m}{2} \right)^2 = 0.03 \, m^2$$

$$2 \times 0.03 \, m^2 = 0.06 \, m^2$$

The area of one 200-mm pipe and one 250-mm pipe (Answer b) is

$$Area \; of \; the \; 250\text{-}mm \; pipe = \pi \left(\frac{0.25 \, m}{2} \right)^2 = 0.05 \, m^2$$

$$0.03 \, m^2 + 0.05 \, m^2 = 0.08 \, m^2$$

The area of one 250-mm pipe and one 150-mm pipe (Answer c) is

$$Area \; of \; the \; 150\text{-}mm \; pipe = \pi \left(\frac{0.15 \, m}{2} \right)^2 = 0.02 \, m^2$$

$$0.05 \, m^2 + 0.02 \, m^2 = 0.07 \, m^2$$

The area of one 250-mm pipe and one 100-mm pipe (Answer d) is

$$Area \; of \; the \; 100\text{-}mm \; pipe = \pi \left(\frac{0.10 \, m}{2} \right)^2 = 0.01 \, m^2$$

$$0.05 \, m^2 + 0.02 \, m^2 = 0.06 \, m^2$$

Correct answer: b

21. Volume = Area × Depth

$$= 37\ m^2 \times 0.9\ m$$

$$= 33.3\ m^3$$

There are 1000 liters in 1 cubic meter.

$$33.3\ \cancel{m^3} \times \frac{1000\ L}{1\ \cancel{m^3}} = 33\ 300\ L$$

Correct answer: b

22. $Slope = \dfrac{Rise}{Run}$

By rearranging the equations, we get

Rise = Slope × Run

$$= 0.026 \times 91\ m$$

$$= 2.366\ rounded\ to\ 2.4\ m$$

Correct answer: b

23. $Velocity = \dfrac{Distance}{Time}$

$$= \frac{82\ m}{75\ s}$$

$$= 1.1\ m/s$$

Correct answer: b

24. Flow = Area × Velocity

There are 1000 millimeters in 1 meter.

$$380\ \cancel{mm} \times \frac{1\ m}{1000\ \cancel{mm}} = 0.38\ m$$

Area of a 380-mm sewer pipe flowing half full $= \dfrac{1}{2}\left[\pi\left(\dfrac{Diam}{2}\right)^2\right]$

$$Area = \frac{1}{2}\pi\left(\frac{0.38\ m}{2}\right)^2 = 0.06\ m^2$$

Flow = 0.06 m² × 0.6 m/s

= 0.36 m³/s

There are 1000 liters per cubic meter.

There are 60 seconds per minute.

$$\frac{0.36\ \cancel{m^3}}{\cancel{s}} \times \frac{1000\ L}{\cancel{m^3}} \times \frac{60\ \cancel{s}}{min} = 2160\ L/min$$

Correct answer: c

25. Volume = Flow × Time

Time = 60 min + 40 min = 100 min

$$Volume = 208\ \frac{L}{\cancel{min}} \times 100\ \cancel{min}$$

= 20 800 L

Correct answer: d

26. $$Pump\ efficiency = \frac{Output,\ kW}{Input\ (brake),\ kW} \times 100$$

$$= \frac{1.32\ \cancel{kW}}{1.87\ \cancel{kW}} \times 100$$

= 70.6% rounded to 71%

Correct answer: c

27. Volume = Length × Width × Depth

= 306 m × 0.77 m × 2.4 m

= 565.49 m³ rounded to 566 m³

Correct answer: b

28. Slope $= \dfrac{\text{Rise}}{\text{Run}}$

Rise $= 62.5 \text{ m} - 62.1 \text{ m}$

Slope $= \dfrac{0.4 \, \cancel{\text{m}}}{99 \, \cancel{\text{m}}} = 0.004$

Correct answer: b

29. Concentration $= \dfrac{\text{Quantity applied}}{\text{Flow}}$

$$= \dfrac{72.6 \, \cancel{\text{kg/d}}}{11\,355 \, \dfrac{\cancel{\text{m}^3}}{\cancel{\text{d}}} \times 0.001 \dfrac{\text{kg/}\cancel{\text{m}^3}}{\text{mg/L}}}$$

$$= 6.4 \text{ mg/L}$$

Correct answer: c

30. $4353 \dfrac{\cancel{\text{L}}}{\cancel{\text{min}}} \times \dfrac{60 \, \cancel{\text{min}}}{\cancel{\text{h}}} \times \dfrac{24 \, \cancel{\text{h}}}{\text{d}} \times \dfrac{1 \text{ m}^3}{1000 \, \cancel{\text{L}}} = 6268 \text{ m}^3/\text{d}$ rounded to $6300 \text{ m}^3/\text{d}$

Correct answer: d

31. Cross-sectional area of a 200-mm pipe $= \pi \left(\dfrac{\text{Diam}}{2} \right)^2 = \pi \left(\dfrac{200 \text{ mm}}{2} \right)^2$

$$= 31\,416 \text{ mm}^2$$

Cross-sectional area of a 150-mm pipe $= \pi \left(\dfrac{\text{Diam}}{2} \right)^2 = \pi \left(\dfrac{150 \text{ mm}}{2} \right)^2$

$$= 17\,671 \text{ mm}^2$$

$$= \dfrac{17\,671 \, \cancel{\text{mm}^2}}{31\,416 \, \cancel{\text{mm}^2}} = 0.56$$

Correct answer: b

32. Volume = Pump rate × Time

$$\text{Time} = 8 \; \cancel{h} \times \left(\frac{60 \; \text{min}}{\cancel{h}} \right) + 47 \; \text{min}$$

$$= 527 \; \text{min}$$

$$\text{Volume} = 1703 \frac{L}{\cancel{\text{min}}} \times 527 \; \cancel{\text{min}}$$

$$= 897\ 481 \; \text{rounded to } 897\ 480 \; L$$

Correct answer: b

33. A. Pipe cost = Linear meters × Cost per linear meter

$$= 259.1 \; \text{m} \times \$12.30/\text{m}$$

$$= \$3,186.93 \; \text{rounded to } \$3,190$$

Correct answer: c

B. Excavation, backfill, and compaction cost = Linear meters × Cost per linear meter

$$= 259.1 \; \cancel{\text{m}} \times \$50.85/\cancel{\text{m}}$$

$$= \$13,175.24 \; \text{rounded to } \$13,175$$

Correct answer: c

C. Cost of replacing the asphalt = Linear meters × Cost per linear meter

$$= 259.1 \; \cancel{\text{m}} \times \$10.50/\cancel{\text{m}}$$

$$= \$2,720.55 \; \text{rounded to } \$2,720$$

Correct answer: a

D. Total cost for the job = Pipe cost + Excavation, backfill, and compaction cost + Asphalt replacement cost

$$= \$3,190 + \$13,175 + \$2,720 = \$19,085$$

Correct answer: c

E. Total cost per linear meter = Total cost for the job/Linear meters

$$= \frac{\$19{,}085}{259.1 \text{ m}}$$

$$= \$73.66/\text{m}$$

Correct answer: c

34.   Flowrate into wet well = Discharge rate + $\dfrac{\text{Change in volume}}{\text{Time}}$

Change in volume = Length $\times$ Width $\times$ Depth

$$= 1.8 \text{ m} \times 1.5 \text{ m} \times 0.3 \text{ m}$$

$$= 0.8 \text{ m}^3$$

There are 1000 liters per cubic meter.

$$0.8 \; \cancel{\text{m}^3} \times \frac{1000 \text{ L}}{\cancel{\text{m}^3}} = 800 \text{ L}$$

$$\text{Time} = 3 \text{ min} + 15 \; \cancel{\text{s}} \times \frac{1 \text{ min}}{60 \; \cancel{\text{s}}} = 3.25 \text{ min}$$

$$\text{Flowrate into wet well} = 1060 \frac{\text{L}}{\text{min}} + \frac{800 \text{ L}}{3.25 \text{ min}}$$

$$= 1306.2 \text{ L/min rounded to } 1306 \text{ L/min}$$

Correct answer: d

35.   Downstream elevation = Upstream elevation − (Slope $\times$ Distance between manholes)

Distance between manholes = Manhole B station − Manhole A station

$$= 220 \text{ m} - 145 \text{ m}$$

$$= 75 \text{ m}$$

Downstream elevation = 43.65 m − (0.0036 $\times$ 75 m)

$$= 43.65 \text{ m} - 0.27 \text{ m}$$

$$= 43.38 \text{ m}$$

Correct answer: b

36. Volume = Length × Width × Depth

Because the bottom of the tank slopes, we need to calculate the average depth.

$$\text{Average depth} = \frac{4.6\ m - 3.7\ m}{2} + 3.7\ m = 4.15\ m$$

Volume = 15.2 m × 5.5 m × 4.15 m = 346.94 m³ rounded to 347 m³

Correct answer: b

37. Volume of pipe to be filled = Cross-sectional area × Length

$$200\ \cancel{mm}\ \times \frac{1\ m}{1000\ \cancel{mm}} = 0.20\ m$$

$$\text{Cross-sectional area} = \pi \left(\frac{\text{Diam}}{2}\right)^2 = \pi \left(\frac{0.20\ m}{2}\right)^2 = 0.0314\ m^2$$

Volume of pipe to be filled = 0.0314 m² × 81 m

$$= 2.54\ m^3$$

There are 1000 liters per cubic meter.

$$2.54\ \cancel{m^3} \times \frac{1000\ L}{1\ \cancel{m^3}} = 2540\ L$$

$$2540\ \cancel{L} \times 250\ mg/\cancel{L} = 635\ 000\ mg\ \text{needed}$$

There are 1 000 000 milligrams per kilogram.

$$635\ 000\ \cancel{mg} \times \frac{1\ kg}{1000\ 000\ \cancel{mg}} = 0.635\ kg\ \text{of material rounded to } 0.64\ kg$$

Correct answer: c

38. Available chlorine = Total weight × Percent available chlorine

$$= 3.6\ kg \times 0.70$$

$$= 2.5\ kg$$

Correct answer: a

39. There are 1000 liters in 1 cubic meter.

$$113\ \frac{\cancel{L}}{s} \times \frac{1\ m^3}{1000\ \cancel{L}} = 0.113\ m^3/s$$

$$0.113\frac{m^3}{\cancel{s}} \times \frac{60\ \cancel{s}}{\cancel{min}} \times \frac{60\ \cancel{min}}{\cancel{h}} \times \frac{24\ \cancel{h}}{d} = 9763\ m^3/d$$

Correct answer: b

40. $$6\ 434\ 500\frac{L}{\cancel{d}} \times \frac{1\ \cancel{d}}{24\ h} = 268\ 104.2\ L/h$$

$$268\ 104.2\frac{L}{\cancel{h}} \times 7\ \cancel{h} = 1\ 876\ 729\ L$$

Correct answer: a

41. $$272\ 520\frac{L}{\cancel{d}} \times \frac{1\ \cancel{d}}{24\ \cancel{h}} \times \frac{1\ \cancel{h}}{60\ min} = 189.25\ L/min\ \text{rounded to } 189\ L/min$$

Correct answer: d

42. Volume = Area × Depth

$$\text{Area} = \pi\left(\frac{\text{Diam}}{2}\right)^2 = \pi\left(\frac{6\ m}{2}\right)^2 = 28.3\ m^2$$

Volume = 28.3 m² × 3.4 m = 96.2 m³

There are 1000 liters in 1 cubic meter.

$$96.2\ \cancel{m^3} \times \frac{1000\ L}{\cancel{m^3}} = 96\ 200\ L$$

Correct answer: b

43. A. Total project cost = (Linear meters of 200 mm pipe × Cost per meter) + (Linear meters of 250-mm pipe × Cost per meter)

There are 1000 meters in 1 kilometer.

$$1.93\ \cancel{km} \times \frac{1000\ m}{\cancel{km}} = 1930\ m$$

$$0.64\ \cancel{km} \times \frac{1000\ m}{\cancel{km}} = 640\ m$$

$$\text{Total project cost} = \left(1930 \ \text{m} \times \frac{\$205.18}{\text{m}}\right) + \left(640 \ \text{m} \times \frac{\$224.40}{\text{m}}\right)$$

$$= \$396,000 + \$143,616$$

$$= \$539,616$$

Correct answer: c

B. Cost of pipe = 20% of Total cost

$$= 0.20 \times \$539,616$$

$$= \$107,923$$

Correct answer: a

C. Staff cost = 100% − (20% + 70%)

$$= 10\% \text{ of Total cost}$$

$$= 0.10 \times \$539\ 616$$

$$= \$53\ 961.60 \text{ rounded to } \$53\ 962$$

Correct answer: b

D. Backfilling and compaction costs = 70% of Total cost

$$= 0.70 \times \$539\ 616$$

$$= \$377,731.20 \text{ rounded to } \$377,731$$

Correct answer: d

44. The calculations and answers for each part of this question are as follows:

A. First, calculate how many linear meters of each size of pipe are in the collection system.

$$200\text{-mm pipe}: 24.14 \ \text{km} \times \frac{1000 \ \text{m}}{\text{km}} = 24\ 140 \ \text{m}$$

$$300\text{-mm pipe}: 4.83 \ \text{km} \times \frac{1000 \ \text{m}}{\text{km}} = 4830 \ \text{m}$$

$$450\text{-mm pipe}: 3.22 \ \text{km} \times \frac{1000 \ \text{m}}{\text{km}} = 3220 \ \text{m}$$

$$600\text{-mm pipe}: 2.41 \ \cancel{\text{km}} \times \frac{1000 \ \text{m}}{\cancel{\text{km}}} = 2410 \ \text{m}$$

Then multiply it by the unit cost.

$$24 \ 140 \ \cancel{\text{m}} \times \frac{\$24.60}{\cancel{\text{m}}} = \$593,844$$

$$4830 \ \cancel{\text{m}} \times \frac{\$25.40}{\cancel{\text{m}}} = \$122,682$$

$$3220 \ \cancel{\text{m}} \times \frac{\$27.10}{\cancel{\text{m}}} = \$87,262$$

$$2410 \ \cancel{\text{m}} \times \frac{\$29.50}{\cancel{\text{m}}} = \$71,095$$

Total cost for contractor cleaning = $593,844 + $122,682 + $87,262 + $71,095 = $874,883 rounded to $875,000

Correct answer: c

B. Total cost for city to clean system = Staff costs + Equipment costs

$$\text{Staff costs} = \text{Hourly rate} \times \text{Number of hours required}$$

$$\text{Number of hours required} = \frac{\text{Total required length}}{\text{Cleaning rate}}$$

$$200\text{-mm pipe}: \frac{24 \ 140 \ \cancel{\text{m}}}{365 \frac{\cancel{\text{m}}}{\text{d}}} = 66.1 \ \text{d}$$

$$300\text{-mm pipe}: \frac{4830 \ \cancel{\text{m}}}{365 \frac{\cancel{\text{m}}}{\text{d}}} = 13.2 \ \text{d}$$

$$450\text{-mm pipe}: \frac{3220 \ \cancel{\text{m}}}{305 \frac{\cancel{\text{m}}}{\text{d}}} = 10.6 \ \text{d}$$

$$600\text{-mm pipe}: \frac{2410 \ \cancel{\text{m}}}{305 \frac{\cancel{\text{m}}}{\text{d}}} = 7.9 \ \text{d}$$

Number of days required = 66.1 d + 13.2 d + 10.6 d + 7.9 d = 97.8 d

$$\text{Staff costs} = 97.8 \ \cancel{\text{d}} \times \frac{8 \ \cancel{\text{h}}}{\cancel{\text{d}}} \times \frac{\$24.00}{\cancel{\text{h}}} = \$18,778$$

Total cost for city to clean system $= \$250,000 + \$18,778$

$$= \$268,778 \text{ rounded to } \$268,800$$

Correct answer: a

C. $\dfrac{\text{Cost of contract cleaning} - \text{Cost of city conducting cleaning}}{\text{Number of years}}$

$$= \dfrac{\$875,000 - \$268,800}{5 \text{ yr}}$$

$$= \$121,240/\text{yr}$$

Correct answer: b

45. $\dfrac{1}{8} = 0.125$

Correct answer: b

46. One liter of water weighs 1 kilogram.

The truck itself approximately weighs an additional 2268 kg.

Full truck weight = 2268 kg + 5675 kg = 7943 kg

Correct answer: d

47. Percentage $= \dfrac{23 \cancel{\text{ joints}}}{90 \cancel{\text{ joints}}} \times 100\%$

$$= 25.6\%$$

Correct answer: c

48. Monthly power cost = Cost per kWh × kW usage × Run time

$$\text{kW usage} = \dfrac{\text{Motor rating}}{\text{Efficiency}} \times \dfrac{746 \text{ W}}{\text{hp}}$$

$$= \dfrac{50 \cancel{\text{ hp}}}{0.84} \times \dfrac{746 \text{ W}}{\cancel{\text{hp}}}$$

$$\text{Run time} = 18 \dfrac{\text{h}}{\cancel{\text{d}}} \times 30 \cancel{\text{ d}}$$

$$= 540 \text{ h}$$

Monthly power cost $= \$0.04/\cancel{kWh} \times 44.4 \ \cancel{kW} \times 540 \ \cancel{h}$

$\qquad\qquad\qquad = \$959.04$ rounded to $\$960$

Correct answer: a

49. $\text{Time} = \dfrac{\text{Distance}}{\text{Velocity}}$

There are 1000 meters in 1 kilometer.

$$\text{Distance} = 5.636 \ \cancel{km} \times \frac{1000 \ m}{\cancel{km}} = 5636 \ m$$

$$\text{Time} = \frac{5636 \ \cancel{m}}{0.61 \ \dfrac{\cancel{m}}{s}} = 9240 \ s$$

$$9240 \ \cancel{s} \times \frac{1 \ \cancel{min}}{60 \ \cancel{s}} \times \frac{1 \ h}{60 \ \cancel{min}} = 2.6 \ h$$

Correct answer: b

50. $\text{Daily amount of chemical used} = 30 \ 280 \dfrac{\cancel{m^3}}{d} \times 12 \ \cancel{mg/L} \times 0.001 \dfrac{kg/\cancel{m^3}}{\cancel{mg/L}}$

$$= 363.36 \ kg/d$$

$$\text{Daily chemical cost} = 363.36 \frac{\cancel{kg}}{d} \times \frac{\$1.35}{22.68 \ \cancel{kg}}$$

$$= \$21.62/d$$

Correct answer: b

51.   Labor cost = Cost per staff hour × Number of people × Number of hours

$$= \frac{\$12.50}{\text{Staff} \ \cancel{h}} \times 2 \ \text{Staff people} \times 10 \ \cancel{h}$$

$$= \$250$$

Correct answer: b

52.   Volume = Length × Width × Depth

$$= 0.76 \ m \times 1.98 \ m \times 19.8 \ m$$

$$= 29.8 \ m^3 \text{ rounded to } 30 \ m^3$$

Correct answer: a

53. $\text{Velocity} = \dfrac{\text{Flow}}{\text{Area}}$

There are 1000 millimeters in 1 meter.

$$381 \ \cancel{\text{mm}} \times \frac{1 \ \text{m}}{1000 \ \cancel{\text{mm}}} = 0.38 \ \text{m}$$

$$\text{Area} = \pi \left(\frac{\text{Diam}}{2} \right)^2$$

$$= \pi \left(\frac{0.38}{2} \right)^2 = 0.11 \ \text{m}^2$$

$$\text{Velocity} = \frac{10 \ 220 \ \text{m}^3/\text{d}}{0.11 \ \cancel{\text{m}^2}} = 92 \ 909 \ \text{m/d}$$

$$92 \ 909 \frac{\text{m}}{\cancel{\text{d}}} \times \frac{\cancel{\text{d}}}{24 \ \cancel{\text{h}}} \times \frac{\cancel{\text{h}}}{60 \ \cancel{\text{min}}} \times \frac{\cancel{\text{min}}}{60 \ \text{s}} = 1.07 \ \text{m/s rounded to } 1.1 \ \text{m/s}$$

Correct answer: a

Answers for Section 8—Mathematics (U.S. Customary Units)

1. Volume of wet well = Length × Width × Depth
   $$= 15 \text{ ft} \times 12 \text{ ft} \times 10 \text{ ft}$$
   $$= 1800 \text{ cu ft}$$

   There are 7.48 gallons in 1 cubic foot.

   $$1800 \, \cancel{\text{cu ft}} \times 7.48 \frac{\text{gal}}{\cancel{\text{cu ft}}} = 13\,464 \text{ gal rounded to } 13\,500 \text{ gal}$$

   Correct answer: b

2. $\text{Velocity} = \dfrac{\text{Flow}}{\text{Area}}$

   $$\text{Flow} = 0.5 \text{ mgd} = 500\,000 \frac{\text{gal}}{\cancel{\text{d}}} \times \frac{1 \, \cancel{\text{d}}}{24 \, \cancel{\text{hr}}} \times \frac{1 \, \cancel{\text{hr}}}{60 \, \cancel{\text{min}}} \times \frac{1 \, \cancel{\text{min}}}{60 \text{ sec}}$$

   $$= 5.8 \frac{\text{gal}}{\text{sec}}$$

   There are 7.48 gallons in 1 cubic foot.

   $$\text{Flow} = 5.8 \frac{\cancel{\text{gal}}}{\text{sec}} \times \frac{\text{cu ft}}{7.48 \, \cancel{\text{gal}}} = 0.78 \frac{\text{cu ft}}{\text{sec}}$$

   The pipe diameter is 8 in. There are 12 inches in 1 foot.

   $$\text{The pipe diameter} = 8 \, \cancel{\text{in.}} \times \frac{1 \text{ ft}}{12 \, \cancel{\text{in.}}} = 0.67 \text{ ft}$$

   $$\text{Area of the pipe} = \pi \left(\frac{\text{Diam}}{2} \right)^2 = \pi \left(\frac{0.67 \text{ ft}}{2} \right)^2 = 0.35 \text{ sq ft}$$

   $$\text{Velocity} = \frac{\text{Flow}}{\text{Area}} = \frac{0.78 \, \frac{\cancel{\text{cu}} \text{ ft}}{\text{sec}}}{0.35 \, \cancel{\text{sq ft}}} = 2.2 \text{ ft/sec}$$

   Correct answer: b

3. Flow = Area × Velocity

   There are 12 inches in 1 foot.

$$\text{Area of the pipe} = \pi \left(\frac{\text{Diam}}{2} \right)^2 = \pi \left(\frac{1\,\text{ft}}{2} \right)^2 = 0.79\ \text{sq ft}$$

$$= 0.79\ \text{sq ft} \times 2\,\frac{\text{ft}}{\text{sec}}$$

$$= 1.58\ \text{cfs rounded to 1.6 cfs}$$

Correct answer: b

4. Volume = Cross-sectional area × Length

There are 12 inches in 1 foot.

$$\text{Area of the pipe} = \pi \left(\frac{\text{Diam}}{2} \right)^2 = \pi \left(\frac{1\,\text{ft}}{2} \right)^2 = 0.79\ \text{sq ft}$$

$$\text{Volume} = 0.79\ \text{sq ft} \times 100\ \text{ft} = 79\ \text{cu ft}$$

There are 7.48 gallons in 1 cubic foot.

$$79\ \cancel{\text{cu ft}} \times 7.48\,\frac{\text{gal}}{\cancel{\text{cu ft}}} = 590.92\ \text{gal rounded up to 591 gal}$$

Correct answer: b

5. $$\text{Area} = \pi \left(\frac{\text{Diam}}{2} \right)^2$$

$$\text{Area} = \pi \left(\frac{35\,\text{ft}}{2} \right)^2 = 962\ \text{sq ft}$$

Correct answer: c

6. Celsius temperature = (5/9) (Fahrenheit temperature − 32)
$$= (5/9)\,(50 - 32)$$
$$= 10\,°\text{C}$$

Correct answer: b

7. $$\text{Time} = \frac{\text{Volume}}{\text{Flow}}$$

Volume = Length × Width × Depth
$$= 6.0\ \text{ft} \times 4.0\ \text{ft} \times 2.0\ \text{ft}$$
$$= 48\ \text{cu ft}$$

One cubic foot holds 7.48 gallons.

$$48 \ \cancel{\text{cu ft}} \times \frac{7.48 \text{ gal}}{\cancel{\text{cu ft}}} = 359 \text{ gal}$$

$$\text{Time} = \frac{359 \ \cancel{\text{gal}}}{24 \ \cancel{\text{gal}}/\text{min}} = 15 \text{ min}$$

Correct answer: c

8. $\dfrac{127 \text{ ft}}{5 \text{ ft}} = 25.4$ pieces of 5-ft lengths of pipe. Because the pipe must be replaced in full

5-ft lengths, 26 sections of pipe must be replaced.

Correct answer: d

9. $\text{Grade} = \dfrac{\text{Rise}}{\text{Run}}$

$$= \frac{3 \text{ ft}}{300 \text{ ft}} = 0.01 \times 100\% = 1.0\%$$

Correct answer: c

10.   Feed rate = Chlorine dose $\times$ Flowrate

$$= 10 \ \cancel{\text{mg/L}} \times 0.37 \frac{\cancel{\text{mil.gal}}}{\text{d}} \times 8.34 \frac{\text{lb}/\cancel{\text{mil.gal}}}{\cancel{\text{mg/L}}}$$

$$= 30.9 \text{ lb/d rounded to } 31 \text{ lb/d}$$

Correct answer: c

11.   There are 12 inches in 1 foot.

$$6 \ \cancel{\text{in.}} \times \frac{1 \text{ ft}}{12 \ \cancel{\text{in.}}} = 0.5 \text{ ft}$$

15 ft + 0.5 ft = 15.5 ft

Correct answer: b

12.   Discharge per stroke = Area of piston head $\times$ Length of stroke

$$6 \ \cancel{\text{in.}} \times \frac{1 \text{ ft}}{12 \ \cancel{\text{in.}}} = 0.5 \text{ ft}$$

$$\text{Area of piston head} = \pi \left(\frac{\text{Diam}}{2} \right)^2 = \pi \left(\frac{0.5 \text{ ft}}{2} \right)^2 = 0.196 \text{ sq ft}$$

Discharge per stroke = 0.196 sq ft × 0.5 ft = 0.098 cu ft

1 cubic foot holds 7.48 gallons.

$$0.098 \; \cancel{\text{cu ft}} \times \frac{7.48 \text{ gal}}{\cancel{\text{cu ft}}} = 0.733 \text{ gal/stroke}$$

$$0.733 \text{ gal/}\cancel{\text{stroke}} \times 16 \; \cancel{\text{stroke}}/\text{min} = 11.7 \text{ gpm}$$

Correct answer: b

13. Flow = Area × Velocity
$$= (3 \text{ ft} \times 2 \text{ ft}) \times 4 \text{ ft/sec}$$
$$= 24 \text{ cfs}$$

Correct answer: c

14. Each foot of water column is equivalent to 0.433 psig.

$$75 \; \cancel{\text{ft}} \times \frac{0.433 \text{ psig}}{\cancel{\text{ft}}} = 32.5 \text{ psig}$$

Correct answer: c

15. Volume = Length × Width × Depth
$$= 40 \text{ ft} \times 4 \text{ ft} \times 7 \text{ ft}$$
$$= 1120 \text{ cu ft}$$

There are 27 cubic feet per cubic yard.

$$1120 \; \cancel{\text{cu ft}} \times \frac{1 \text{ cu yd}}{27 \; \cancel{\text{cu ft}}} = 41.5 \text{ cu yd}$$

Correct answer: c

16. $$\frac{50 \text{ ft}}{320 \text{ ft}} = \frac{1 \text{ in.}}{x}$$

$$50 \text{ ft } x = 320 \text{ ft} \times 1 \text{ in.}$$

$$x = \frac{320 \text{ ft}}{50 \text{ ft}} \times 1 \text{ in.}$$

$$x = 6.4 \text{ in.}$$

Correct answer: a

17. BOD loading = BOD concentration × Flow

$$= 120 \; \cancel{\text{mg/L}} \times 18 \frac{\cancel{\text{mil. gal}}}{\text{d}} \times 8.34 \frac{\text{lb/}\cancel{\text{mil. gal}}}{\cancel{\text{mg/L}}}$$

$$= 18\,014.4 \; \text{lb/d rounded to } 18\,000 \; \text{lb/d}$$

Correct answer: c

18. Volume = Length × Width × Depth

   = 65 ft × 2.5 ft × 6.5 ft

   = 1056.25 cu ft

There are 27 cubic feet in 1 cubic yard.

$$1056.25 \; \cancel{\text{cu ft}} \times \frac{1 \; \text{cu yd}}{27 \; \cancel{\text{cu ft}}} = 39.1 \; \text{cu yd rounded to } 39 \; \text{cu yd}$$

Correct answer: c

19. Volume = Length × Width × Depth

Our pump turns on at a 9-ft depth and shuts off at a 3-ft depth. Therefore, the depth of water removed during each cycle is

9 ft − 3 ft = 6 ft

   Volume = 10 ft × 10 ft × 6 ft

   = 600 cu ft

There are 7.48 gallons in 1 cubic foot.

$$600 \; \cancel{\text{cu ft}} \times \frac{7.48 \; \text{gal}}{1 \; \cancel{\text{cu ft}}} = 4488 \; \text{gal}$$

Correct answer: b

20. The cross-sectional area of a 12-in. diameter pipe is

$$\text{Area} = \pi r^2 = \pi \left(\frac{\text{Diam}}{2} \right)^2 = \pi \left(\frac{1 \; \text{ft}}{2} \right)^2$$

$$= 0.875 \; \text{sq ft}$$

The area of which combination of smaller pipes is most closely equal to this?

The area of two 8-in. pipes (Answer a) is

$$\text{Area of an 8-in. pipe} = \pi\left(\frac{\text{Diam}}{2}\right) = \pi\left(\frac{0.67\ \text{ft}}{2}\right)^2 = 0.35\ \text{sq ft}$$

$$2 \times 0.35\ \text{sq ft} = 0.7\ \text{sq ft}$$

The area of one 8-in. pipe and one 10-in. pipe (Answer b) is

$$\text{Area of the 10-in. pipe} = \pi\left(\frac{0.83\ \text{ft}}{2}\right)^2 = 0.54\ \text{sq ft}$$

$$0.35\ \text{sq ft} + 0.54\ \text{sq ft} = 0.89\ \text{sq ft}$$

The area of one 10-in. pipe and one 6-in. pipe (Answer c) is

$$\text{Area of the 6-in. pipe} = \pi\left(\frac{0.5\ \text{ft}}{2}\right)^2 = 0.20\ \text{sq ft}$$

$$0.54\ \text{sq ft} + 0.20\ \text{sq ft} = 0.74\ \text{sq ft}$$

The area of one 10-in. pipe and one 4-in. pipe (Answer d) is

$$\text{Area of the 4-in. pipe} = \pi\left(\frac{0.33\ \cancel{\text{ft}}}{2}\right)^2 = 0.09\ \cancel{\text{ft}}$$

$$0.54\ \text{sq ft} + 0.09\ \text{sq ft} = 0.63\ \text{sq ft}$$

Correct answer: c

21. Volume = Area × Depth
    = 400 sq ft × 3 ft
    = 1200 cu ft

There are 7.48 gallons in 1 cubic foot.

$$1200\ \cancel{\text{cu ft}} \times 7.48\frac{\text{gal}}{\cancel{\text{cu ft}}} = 8976\ \text{gal}$$

Correct answer: b

22. $\text{Slope} = \dfrac{\text{Rise}}{\text{Run}}$

By rearranging the equation, we get

Rise = Slope × Run
    = 0.026 × 300 ft
    = 7.8 ft of drop

Correct answer: b

23. $\text{Velocity} = \dfrac{\text{Distance}}{\text{Time}}$

$= \dfrac{268 \text{ ft}}{75 \text{ sec}}$

$= 3.57 \text{ ft/sec}$

Correct answer: b

24. $\text{Flow} = \text{Area} \times \text{Velocity}$

$\text{Area of a 15-in. sewer pipe flowing half full} = \dfrac{1}{2}\left[\pi\left(\dfrac{\text{Diam}}{2}\right)^2\right]$

There are 12 inches in 1 foot.

$15 \; \cancel{\text{in.}} \times \dfrac{1 \text{ ft}}{12 \; \cancel{\text{in.}}} = 1.25 \text{ ft} = \dfrac{1}{2}\left[\pi\left(\dfrac{1.25}{2}\right)^2\right] = 0.61 \text{ sq ft}$

$\text{Flow} = 0.61 \text{ sq ft} \times 2 \text{ ft/sec}$

$= 1.22 \text{ cfs}$

There are 7.48 gallons per cubic foot.

There are 60 seconds per minute.

$1.22 \; \cancel{\text{cu ft}}/\cancel{\text{sec}} \times \dfrac{7.48 \text{ gal}}{\cancel{\text{cu ft}}} \times \dfrac{60 \; \cancel{\text{sec}}}{\text{min}} = 547.5 \text{ gpm rounded to 550 gpm}$

Correct answer: c

25. $\text{Volume} = \text{Flow} \times \text{Time}$

$\text{Time} = 60 \text{ min} + 40 \text{ min} = 100 \text{ min}$

$\text{Volume} = 55 \dfrac{\text{gal}}{\cancel{\text{min}}} \times 100 \; \cancel{\text{min}}$

$= 5500 \text{ gal}$

Correct answer: d

26. Pump efficiency $= \dfrac{\text{Output, hp}}{\text{Input (brake), hp}} \times 100$

$= \dfrac{1.77 \text{ hp}}{2.50 \text{ hp}} \times 100$

$= 70.8\%$ rounded to 71%

Correct answer: c

27. Volume = Length $\times$ Width $\times$ Depth

$= 1000 \text{ ft} \times 2.5 \text{ ft} \times 8 \text{ ft}$

$= 20\,000 \text{ cu ft}$

There are 27 cubic feet in each cubic yard.

$20\,000 \text{ cu ft} \times \dfrac{1 \text{ cu yd}}{27 \text{ cu ft}} = 740.7 \text{ cu yd}$ rounded to 741 cu yd

Correct answer: b

28. Slope $= \dfrac{\text{Rise}}{\text{Run}}$

Rise $= 205 \text{ ft} - 203.7 \text{ ft}$

Slope $= \dfrac{1.3 \text{ ft}}{325 \text{ ft}} = 0.004$

Correct answer: b

29. Concentration $= \dfrac{\text{Quantity applied}}{\text{Flow}}$

$= \dfrac{160 \text{ lb/d}}{3 \text{ mil. gal/d} \times 8.34 \dfrac{\text{lb/mil. gal}}{\text{mg/L}}}$

$= 6.4 \text{ mg/L}$

Correct answer: c

30. $1150\dfrac{\cancel{gal}}{\cancel{min}}\times\dfrac{60\ \cancel{min}}{\cancel{hr}}\times\dfrac{24\ \cancel{hr}}{d}\times\dfrac{1\ \text{mil. gal}}{1\ 000\ 000\ \cancel{gal}}=1.66$ mgd rounded to 1.7 mgd

Correct answer: d

31.   Cross-sectional area of an 8-in. pipe $=\pi\left(\dfrac{\text{Diam}}{2}\right)^2=\pi\left(\dfrac{8}{2}\right)^2$

$$=50.3\ \text{in.}^2$$

Cross-sectional area of a 6-in. pipe $=\pi\left(\dfrac{\text{Diam}}{2}\right)^2=\pi\left(\dfrac{6}{2}\right)^2$

$$=28.3\ \text{in.}^2$$

$$\dfrac{28.3\ \text{in.}^2}{50.3\ \text{in.}^2}=0.56$$

Correct answer: b

32.   Volume $=$ Pump rate $\times$ Time

$$\text{Time}=\left(8\ \cancel{hr}\times\dfrac{60\ \text{min}}{\cancel{hr}}\right)+47\ \text{min}$$

$$=527\ \text{min}$$

$$\text{Volume}=450\dfrac{\text{gal}}{\cancel{min}}\times527\ \cancel{min}$$

$$=237\ 150\ \text{gal rounded to }237\ 000\ \text{gal}$$

Correct answer: b

33.   A. Pipe cost $=$ Linear footage $\times$ Cost per linear foot
$$= 850\ \cancel{ft}\times \$3.75/\cancel{ft}$$
$$= \$3,187.50\text{ rounded to }\$3,190$$

Correct answer: c

B. Excavation, backfill and compaction cost $=$ Linear footage $\times$ Cost per linear foot
$$= 850\ \cancel{ft}\times \$15.50/\cancel{ft}$$
$$= \$13,175$$

Correct answer: c

C. Cost of replacing the asphalt = Linear footage × Cost per linear foot

$$= 850 \text{ ft} \times \$3.20 \text{ ft}$$

$$= \$2,720$$

Correct answer: a

D. Total cost for the job = Pipe cost + Excavation, backfill, and compaction cost + Asphalt replacement cost

$$= \$3,190 + \$13,175 + \$2,720 = \$19,085$$

Correct answer: c

E. Total cost per linear foot = Total cost for the job/Linear footage

$$= \frac{\$19,085}{850 \text{ ft}}$$

$$= \$22.45/\text{ft}$$

Correct answer: c

34. $\text{Flowrate into wet well} = \text{Discharge rate} + \dfrac{\text{Change in volume}}{\text{Time}}$

Change in volume = Length × Width × Depth

$$= 6 \text{ ft} \times 5 \text{ ft} \times 1 \text{ ft}$$

$$= 30 \text{ cu ft}$$

There are 7.48 gallons per cubic foot.

$$30 \text{ cu ft} \times \frac{7.48 \text{ gal}}{\text{cu ft}} = 224.4 \text{ gal}$$

$$\text{Time} = 3 \text{ min} + \left(15 \text{ sec} \times \frac{1 \text{ min}}{60 \text{ sec}}\right) = 3.25 \text{ min}$$

$$\text{Flowrate into wet well} = 280 \frac{\text{gal}}{\text{min}} + \left(\frac{224.4 \text{ gal}}{3.25 \text{ min}}\right)$$

$$= 349 \text{ gpm}$$

Correct answer: d

35. Downstream elevation = Upstream elevation − (Slope × Distance between manholes)

Distance between manholes = Manhole B station − Manhole A station

$$= 721.7 \text{ ft} - 475.5 \text{ ft}$$

$$= 246.2 \text{ ft}$$

Downstream elevation = 143.21 ft − (0.0036 × 246.2 ft)

$$= 143.21 \text{ ft} - 0.89 \text{ ft}$$

$$= 142.32 \text{ ft}$$

Correct answer: b

36. Volume = Length × Width × Depth

Because the bottom of the tank slopes, we need to calculate the average depth.

$$\text{Average depth} = \frac{15-12}{2} + 12 = 13.5 \text{ ft}$$

Volume = 50 ft × 18 ft × 13.5 ft = 12 150 cu ft

Correct answer: b

37. Volume of pipe to be filled = Cross-sectional area × Length

$$18 \text{ in.} \times \frac{1 \text{ ft}}{12 \text{ in.}} = 0.67 \text{ ft}$$

$$\text{Cross-sectional area} = \pi \left(\frac{\text{Diam}}{2} \right)^2 = \pi \left(\frac{0.67}{2} \right)^2 = 0.35 \text{ sq ft}$$

Volume of pipe to be filled = 0.35 sq ft × 265 ft

$$= 92.75 \text{ cu ft}$$

There are 7.48 gallons per cubic foot.

$$92.75 \text{ cu ft} \times \frac{7.48 \text{ gal}}{\text{cu ft}} = 693.77 \text{ gal}$$

There are 3.785 liters per gallon.

$$693.77 \text{ gal} \times \frac{3.785 \text{ L}}{\text{gal}} = 2625.92 \text{ L}$$

$$2625.92 \text{ L} \times 250 \text{ mg/L} = 656 \ 480 \text{ mg needed}$$

There are 1 000 000 milligrams in 1 kilogram.

There are 2.2 pounds in 1 kilogram.

$$656\,480 \text{ mg} \times \frac{1 \text{ kg}}{1\,000\,000 \text{ mg}} \times \frac{2.2 \text{ lb}}{1 \text{ kg}} = 1.44 \text{ lb}$$

Correct answer: c

38. Available chlorine = Total weight × Percent available chlorine

$$= 8 \text{ lb} \times 0.70$$

$$= 5.6 \text{ lb}$$

Correct answer: a

39. There are 7.48 gallons per cubic foot.

$$4 \frac{\text{cu ft}}{\text{sec}} \times \frac{7.48 \text{ gal}}{\text{cu ft}} = 29.92 \text{ gal/sec}$$

$$29.92 \frac{\text{gal}}{\text{sec}} \times \frac{60 \text{ sec}}{\text{min}} \times \frac{60 \text{ min}}{\text{hr}} \times \frac{24 \text{ hr}}{\text{d}} = 2\,585\,088 \text{ gpd}$$

$$\frac{2\,585\,088 \frac{\text{gal}}{\text{d}}}{1\,000\,000 \frac{\text{gal}}{\text{mil. gal}}} = 2.59 \text{ mgd}$$

Correct answer: b

40. $$1\,700\,000 \frac{\text{gal}}{\text{d}} \times \frac{1 \text{ d}}{24 \text{ hr}} = 70\,833 \text{ gph}$$

$$70\,833 \frac{\text{gal}}{\text{d}} \times 7 \text{ hr} = 495\,831 \text{ gal}$$

Correct answer: a

41. $$72\,000 \frac{\text{gal}}{\text{d}} \times \frac{1 \text{ d}}{24 \text{ hr}} \times \frac{1 \text{ hr}}{60 \text{ min}} = 50 \text{ gpm}$$

Correct answer: d

42. Volume = Area × Depth

$$\text{Area} = \pi \left(\frac{\text{Diam}}{2} \right)^2 = \pi \left(\frac{20 \text{ ft}}{2} \right)^2 = 314.2 \text{ sq ft}$$

Volume = 314.2 sq ft × 11 ft = 3456 cu ft

There are 7.48 gallons per cubic foot.

$$3456 \text{ cu ft} \times \frac{7.48 \text{ gal}}{\text{cu ft}} = 25\,851 \text{ gal}$$

Correct answer: b

43. The calculations and answers for each part of this question are as follows:

A. Total project cost = (Linear footage of 8-in. pipe × Cost per linear foot) +
            (Linear footage of 10-in. pipe × Cost per linear foot)

There are 5280 feet in 1 mile.

$$1.2 \text{ mile} \times \frac{5280 \text{ ft}}{\text{mile}} = 6336 \text{ ft}$$

$$0.4 \text{ mile} \times \frac{5280 \text{ ft}}{\text{mile}} = 2112 \text{ ft}$$

$$\text{Total project cost} = (6336 \text{ ft} \times \$62.50/\text{ft}) + (2112 \text{ ft} \times \$68.00/\text{ft})$$

$$= \$396,000 + \$143,616$$

$$= \$539,616$$

Correct answer: c

B. Cost of pipe = 20% of Total cost

$$= 0.20 \times \$539,616$$

$$= \$107,923$$

Correct answer: a

C. Staff cost = 100% − (20% + 70%)

$$= 10\% \text{ of Total cost}$$

$$= 0.10 \times \$539,616$$

$$= \$53,961.60 \text{ rounded to } \$53,962$$

Correct answer: b

D. Backfilling and compaction costs = 70% of Total cost

$$= 0.70 \times \$539{,}616$$

$$= \$377{,}731.20 \text{ rounded to } \$377{,}731$$

Correct answer: d

44. The calculations and answers for each part of this question are as follows:

A. First, calculate how many linear feet of each size of pipe are in the collection system.

$$8\text{-in. pipe}: 15 \;\cancel{\text{mile}} \times \frac{5280 \text{ ft}}{1 \;\cancel{\text{mile}}} = 79\;200 \text{ ft}$$

$$12\text{-in. pipe}: 3 \;\cancel{\text{mile}} \times \frac{5280 \text{ ft}}{1 \;\cancel{\text{mile}}} = 15\;840 \text{ ft}$$

$$18\text{-in. pipe}: 2 \;\cancel{\text{mile}} \times \frac{5280 \text{ ft}}{1 \;\cancel{\text{mile}}} = 10\;560 \text{ ft}$$

$$24\text{-in. pipe}: 1.5 \;\cancel{\text{mile}} \times \frac{5280 \text{ ft}}{1 \;\cancel{\text{mile}}} = 7920 \text{ ft}$$

Then, multiply it by the unit cost.

$$79\;200 \;\cancel{\text{ft}} \times \frac{\$7.50}{\cancel{\text{ft}}} = \$594{,}000$$

$$15\;840 \;\cancel{\text{ft}} \times \frac{\$7.75}{\cancel{\text{ft}}} = \$122{,}760$$

$$10\;560 \;\cancel{\text{ft}} \times \frac{\$8.25}{\cancel{\text{ft}}} = \$87{,}120$$

$$7920 \;\cancel{\text{ft}} \times \frac{\$9.00}{\cancel{\text{ft}}} = \$71{,}280$$

Total cost for contractor cleaning = \$594,000 + \$122,760 + \$87,120 + \$71,280

$$= \$875{,}160 \text{ rounded to } \$875{,}000$$

Correct answer: c

B. Total cost for city to clean system = Staff costs + Equipment costs

Staff costs = Hourly rate × Number of hours required

$$\text{Number of hours required} = \frac{\text{Total linear footage}}{\text{Cleaning rate}}$$

$$\text{8-in. pipe}: \frac{79\ 200\ \cancel{\text{ft}}}{1200\ \dfrac{\cancel{\text{ft}}}{\text{d}}} = 66\ \text{d}$$

$$\text{12-in. pipe}: \frac{15\ 840\ \cancel{\text{ft}}}{1200\ \dfrac{\cancel{\text{ft}}}{\text{d}}} = 13.2\ \text{d}$$

$$\text{18-in. pipe}: \frac{10\ 560\ \cancel{\text{ft}}}{1000\ \dfrac{\cancel{\text{ft}}}{\text{d}}} = 10.6\ \text{d}$$

$$\text{24-in. pipe}: \frac{7920\ \cancel{\text{ft}}}{1000\ \dfrac{\cancel{\text{ft}}}{\text{d}}} = 7.9\ \text{d}$$

$$\text{Number of days required} = 66\ \text{d} + 13.2\ \text{d} + 10.6\ \text{d} + 7.9\ \text{d} = 97.7\ \text{d}$$

$$\text{Staff costs} = 97.7\ \cancel{\text{d}} \times \frac{8\ \cancel{\text{hr}}}{\cancel{\text{d}}} \times \frac{\$24.00}{\cancel{\text{hr}}} = \$18,758.40$$

$$\text{Total cost for city to clean system} = \$250,000.00 + \$18,758.40$$

$$= \$268,758.40 \text{ rounded to } \$268,800$$

Correct answer: a

$$\text{C.} \quad \frac{\text{Cost of contract cleaning} - \text{Cost of city conducting the cleaning}}{\text{Number of years}}$$

$$= \frac{\$875,000 - \$268,800}{5 \text{ years}}$$

$$= \$121,240/\text{yr}$$

Correct answer: b

45. $\dfrac{1}{8} = 0.125$

Correct answer: b

46. Water weighs 8.34 pounds per gallon.

$$\text{Water weight} = 1500 \text{ gal} \times 8.34 \frac{\text{lb}}{\text{gal}}$$

$$= 12\ 510\ \text{lb}$$

The truck itself approximately weighs an additional 5000 lb.

Full truck weight = 5000 lb + 12 510 lb

$$= 17\ 510\ \text{lb rounded to } 17\ 500\ \text{lb}$$

Correct answer: d

47. $$\text{Percentage} = \frac{23 \text{ joints}}{90 \text{ joints}} \times 100\%$$

$$= 25.6\%$$

Correct answer: c

48. Monthly power cost = Cost per kWh × kW usage × Run time

$$\text{kW usage} = \frac{\text{Motor rating}}{\text{Efficiency}} \times \frac{746\ \text{W}}{\text{hp}}$$

$$= \frac{50 \text{ hp}}{0.84} \times \frac{746\ \text{W}}{\text{hp}}$$

$$= 44\ 405 \text{ W} \times \frac{1\ \text{kW}}{1000 \text{ W}} = 44.4\ \text{kW}$$

$$\text{Run time} = 18 \frac{\text{hr}}{\text{d}} \times 30 \text{ d}$$

$$= 540\ \text{hr}$$

Monthly power cost = $0.04/\text{kWh} \times 44.4 \text{ kW} \times 540 \text{ hr}$

$$= \$959.04 \text{ rounded to } \$960$$

Correct answer: a

49. $$\text{Time} = \frac{\text{Distance}}{\text{Velocity}}$$

There are 5280 feet in 1 mile.

$$\text{Distance} = 3.5 \; \cancel{\text{mile}} \times \frac{5280 \; \text{ft}}{\cancel{\text{mile}}}$$

$$= 18 \; 480 \; \text{ft}$$

$$\text{Time} = \frac{18 \; 480 \; \cancel{\text{ft}}}{2 \dfrac{\cancel{\text{ft}}}{\text{sec}}} = 9240 \; \text{sec}$$

There are 60 seconds in 1 minute.

There are 60 minutes in 1 hour.

$$9240 \; \cancel{\text{sec}} \times \frac{1 \; \cancel{\text{min}}}{60 \; \cancel{\text{sec}}} \times \frac{1 \; \text{hr}}{60 \; \cancel{\text{min}}} = 2.6 \; \text{hr}$$

Correct answer: b

50. Daily amount of chemical used $= 8 \dfrac{\cancel{\text{mil. gal}}}{\text{d}} \times 12 \; \cancel{\text{mg/L}} \times 8.34 \dfrac{\text{lb/} \cancel{\text{mil. gal}}}{\cancel{\text{mg/L}}}$

$$= 800.64 \; \text{lb/d}$$

$$\text{Daily chemical cost} = 800.64 \dfrac{\cancel{\text{lb}}}{\text{d}} \times \frac{\$1.35}{50 \; \cancel{\text{lb}}}$$

$$= \$21.62/\text{d}$$

Correct answer: b

51. Labor cost = Cost per staff hour × Number of people × Number of hours

$$= \frac{\$12.50}{\text{Staff} \; \cancel{\text{hr}}} \times 2 \; \text{Staff people} \times 10 \; \cancel{\text{hr}}$$

$$= \$250$$

Correct answer: b

52. Volume = Length × Width × Depth

$$= 65 \; \text{ft} \times 2.5 \; \text{ft} \times 6.5 \; \text{ft}$$

$$= 1056.25 \; \text{cu ft}$$

There are 27 cubic feet in 1 cubic yard.

$$1056.25 \; \text{cu ft} \times \frac{1 \; \text{cu yd}}{27 \; \text{cu ft}} = 39.1 \; \text{cu yd rounded to 39 cu yd}$$

Correct answer: a

53. $\text{Velocity} = \dfrac{\text{Flow}}{\text{Area}}$

$\text{Flow} = 2.7\,\dfrac{\text{mil. gal}}{\text{d}} = 2\,700\,000\ \text{gpd}$

There are 7.48 gallons in 1 cubic foot.

$\dfrac{2\,700\,000\ \dfrac{\cancel{\text{gal}}}{\text{d}}}{7.48\ \dfrac{\cancel{\text{gal}}}{\cancel{\text{cu ft}}}} = 360\,963\ \text{cu ft/d}$

There are 12 inches in 1 foot.

$\text{Diameter} = 15\ \cancel{\text{in.}} \times \dfrac{1\ \text{ft}}{12\ \cancel{\text{in.}}} = 1.25\ \text{ft}$

$\text{Area} = \pi \left(\dfrac{\text{Diam}}{2}\right)^2 = \pi \left(\dfrac{1.25\ \text{ft}}{2}\right)^2 = 1.23\ \text{sq ft}$

$\text{Velocity} = \dfrac{360\,963\ \dfrac{\cancel{\text{cu ft}}}{\text{d}}}{1.23\ \cancel{\text{sq ft}}} = 293\,466\ \text{ft/d}$

$= 293\,466\,\dfrac{\text{ft}}{\cancel{\text{d}}} \times \dfrac{\cancel{\text{d}}}{24\ \cancel{\text{hr}}} \times \dfrac{\cancel{\text{hr}}}{60\ \cancel{\text{min}}} \times \dfrac{\cancel{\text{min}}}{60\ \text{sec}} = 3.4\ \text{ft/sec}$

Correct answer: a

REFERENCES

American Society of Civil Engineers; Water Environment Federation (1991) *Design and Construction of Sanitary and Storm Sewers*, 6th ed.; WEF Manual of Practice No. 9; American Society of Civil Engineers: Reston, Virginia.

American Society of Civil Engineers; Water Environment Federation (1992) *Design and Construction of Urban Stormwater Management Systems*; WEF Manual of Practice No. FD-20; American Society of Civil Engineers: Reston, Virginia.

American Society of Civil Engineers; Water Environment Federation (2007) *Gravity Sanitary Sewer Design and Construction*; WEF Manual of Practice No. FD-5; American Society of Civil Engineers : Reston, Virginia.

California State University (2018) *Operation and Maintenance of Wastewater Collection Systems*, 8th ed.; California State University: Sacramento, California.

Occupational Safety and Health Administration (1996) Occupational Safety and Health Standards; Permit-required confined spaces; *29 CFR 1910.146;* U.S. Department of Labor: Washington, D.C.

Price, J. K. (1991) *Applied Math for Wastewater Plant Operators;* CRC Press: Boca Raton, Florida.

Water Environment Federation; American Society of Civil Engineers (2009) *Existing Sewer Evaluation and Rehabilitation*, 3rd ed.; WEF Manual of Practice No. FD-6; Water Environment Federation: Alexandria, Virginia.

Water Environment Federation (1993) *Design of Wastewater and Stormwater Pumping Stations*, 2nd ed.; Manual of Practice FD-4; Water Environment Federation: Alexandria, Virginia.

Water Environment Federation (2016) *Operation of Water Resource Recovery Facilities*, 7th ed.; Manual of Practice No. 11; Water Environment Federation: Alexandria, Virginia.

Water Environment Federation (2011) *Prevention and Control of Sewer System Overflows*, 3rd ed.; Manual of Practice No. FD-17; Water Environment Federation: Alexandria, Virginia.

Water Environment Federation (2012) *Safety, Health, and Security in Wastewater Systems*, 6th ed.; Manual of Practice No. 1; Water Environment Federation: Alexandria, Virginia.

Water Environment Federation (2009) *Wastewater Collection Systems Management*, 6th ed.; Manual of Practice No. 7; Water Environment Federation: Alexandria, Virginia.

GLOSSARY

abrasion-resistant material   Material that is hard and resistant to physical wear caused by friction.

acid   (1) A substance that tends to lose a proton. (2) A substance that dissolves in water with the formation of hydrogen ions. (3) A substance containing hydrogen, which may be replaced by metals to form salts.

acid-forming bacteria   Microorganisms that can metabolize complex organic compounds under anaerobic conditions. This metabolic activity is the first step in the two-step anaerobic fermentation process leading to the production of methane.

acidity   The quantitative capacity of aqueous solutions to neutralize a base; measured by titration with a standard solution of a base to a specified endpoint; typically expressed as milligrams of equivalent calcium carbonate per liter (mg/L $CaCO_3$); not to be confused with pH. Water does not have to have a low pH to have high acidity.

acre-foot (ac-ft)   A volume that is 1 ft deep and 1 ac in area, or 43 560 cu ft (1 233.5 m³).

aeration tank   A tank in which air is added to wastewater to create optimal conditions to cultivate the growth of healthy biomass.

aerobic   Requiring, or not destroyed by, the presence of free or dissolved oxygen in an aqueous environment.

aerobic bacteria   Bacteria that require free elemental oxygen to sustain life.

air binding   The reduction in fluid flowrate through a structure or pipe caused by the increasing pressure of entrapped air.

air-bound   Obstructed, as to the free flow of water, because of air entrapped in a high point; used to describe a pipeline or pump in such a condition.

air-chamber pump   A displacement pump equipped with an air chamber in which the air is alternately compressed and expanded by the water displaced by the pump, resulting in the water being discharged at a more even rate.

air change   A performance measurement of a ventilation system that refers to the rate at which the system can remove the total volume of air in an enclosed space and replace it with ambient air.

air gap   The unobstructed vertical distance through the free atmosphere between the lowest opening from any pipe or outlet supplying water to a tank, plumbing fixture, or other device, and the flood-level rim of the receptacle.

air injector   A component of an air-lift pump made up of an air supply line and diffuser, which introduces air to the vertical eductor pipe.

air lift   A device for raising liquid by injecting air in and near the bottom of a riser pipe submerged in the liquid to be raised.

air-lift pump   A vertical eductor pipe with an air injector submerged in the fluid to be pumped. Air bubbles are introduced to the bottom of the eductor so that the denser water surrounding it pushes the aerated water up through the discharge pipe. These pumps can be used to pump smaller flows of untreated wastewater or return sludge.

air lock   A reduction or stoppage of liquid flow caused by entrapped air. See also *air binding*.

air-relief valve   A valve that releases air automatically without losing water or that lets air into a line if the internal pressure falls below atmospheric pressure.

air supply valve   A valve used to throttle or isolate an air supply line.

air-powered diaphragm pump   A type of reciprocating pump in which compressed air flexes a membrane that is pushed or pulled to contract or enlarge an enclosed cavity. The most common application is to pump sludge from primary sedimentation tanks and gravity thickeners.

algae   Photosynthetic microscopic plants that contain chlorophyll that float or are suspended in water. They may also be attached to structures, rocks, and so on. In high concentrations, algae may deplete dissolved oxygen in receiving waters.

alkali   Generally, any substance that has highly basic properties; used particularly in reference to the soluble salts of sodium, potassium, calcium, and magnesium.

alkaline   The condition of water, wastewater, or soil that contains a sufficient amount of alkali substances to raise the pH above 7.0.

alkalinity   The capacity of water to neutralize acids; a property imparted by carbonates, bicarbonates, hydroxides, and occasionally borates, silicates, and phosphates. It is expressed in milligrams of equivalent calcium carbonate per liter (mg/L CaCO3).

alternating current   An electric current that reverses its direction at regular intervals.

ambient   Generally refers to the prevailing dynamic environmental conditions in a given area.

ammeter   An instrument for measuring electric current, either alternating or direct, in units of amperes.

ammonia nitrogen   The quantity of elemental nitrogen present in the form of ammonia (NH_3).

ampere   Standard unit of electric current measurement. One ampere represents the flow of one coulomb of electricity per second. A flow of one ampere is produced in a resistance of one ohm by a potential difference of 1 V.

amperometric   Pertaining to measurement of electric current flowing or generated, rather than by voltage.

anaerobic   (1) A condition in which free and dissolved oxygen is unavailable. (2) Requiring or not destroyed by the absence of air or free oxygen.

anaerobic bacteria   Bacteria that grow only in the absence of free and dissolved oxygen.

anaerobic digestion   The degradation of concentrated wastewater solids, during which anaerobic bacteria break down the organic material into inert solids, water, carbon dioxide, and methane.

angular misalignment   A type of shaft misalignment that results from the pump shaft and the motor shaft coming together without having parallel axial alignment.

anti-siphoning spring   A spring added to discharge valves in systems where pumps are located below the liquid level in the suction tank. The spring pushes the discharge valve closed when the pump is off to prevent fluid from moving through the system by siphoning action.

appurtenances   Machinery, appliances, or auxiliary structures attached to a main structure enabling it to function, but not considered an integral part of it.

area drain   A drain installed to collect surface or stormwater from an open area of a building.

armature   Rotating member of a brushed direct current motor in which electromotive force is produced by magnetic induction or a stationary component with same function in an alternating current motor.

automatic oiler   A type of oiler that automatically regulates the oil reservoir level to keep a bearing properly lubricated.

automatic timer   A feature of some pump systems that starts the pumps at regular intervals for a preset length of time. These systems are frequently used for sludge pumping.

average   An arithmetic mean obtained by adding quantities and dividing the sum by the number of quantities.

average daily flow   (1) The total quantity of liquid tributary to a point divided by the number of days of flow measurement. (2) In water and wastewater applications, the total flow past a point over a period of time divided by the number of days in that period.

average flow   Arithmetic average of flows measured at a given point.

average velocity   The average velocity of a stream flowing in a channel or conduit at a given cross section or in a given reach. It is equal to the discharge divided by the cross-sectional area of the section or the average cross-sectional area of the reach. Also called *mean velocity*.

axial flow impeller   Impellers that resemble propellers. Pumps incorporating this type of impeller are used for pumping treated effluent or clean water. They are less useful for raw wastewater or sludge because the tight tolerances of the impellers cannot easily handle solids or stringy material.

babbitt bearing   A type of bearing lined with a soft alloy containing tin, copper, and antimony. The function of the alloy is to reduce friction.

backflow connection   In plumbing, any arrangement whereby backflow can occur. Also called *interconnection*. See also *cross-connection*.

backflow preventer   A device on a water supply pipe to prevent the backflow of water into the water supply system from the connections on its outlet end. See also *vacuum breaker* and *air gap*.

backflow prevention system   A system of check valves and pressure relief valves used to prevent the flow of water through a line in a direction opposite to normal or intended flow. A typical system consists of two spring-loaded check valves in series separated by a pressure relief valve. Also, any effective device, method, or construction used to prevent backflow into a potable water system.

backflushing   The action of reversing the flow through a conduit for the purpose of cleaning the conduit of deposits.

back-pressure valve   A valve provided with a disk that is hinged on the edge so that it opens in the direction of normal flow and closes with reverse flow; a check valve.

bacteria   A group of universally distributed, rigid, essentially unicellular microscopic organisms lacking chlorophyll. These organisms perform a variety of biological treatment processes, including biological oxidation, sludge digestion, nitrification, and denitrification.

ball bearing   A component in machinery in which the journal turns upon loose hardened steel balls that roll easily in a grooved track so that friction is reduced; also: one of the balls in such a bearing.

ball check valve   A non-return valve in which a ball sits within a cylindrical fluid line. It is designed to prevent backflow through the line in which it is installed.

bar screen   A screen composed of parallel bars, either vertical or inclined, placed in a waterway to catch debris. The debris is raked from the screen manually or automatically. Also called a bar rack.

barminutor   A bar screen of standard design fitted with an electrically operated shredding device that sweeps vertically up and down the screen cutting up material retained on the screen.

base   A compound that dissociates in aqueous solution to yield hydroxyl ions.

basic data   Records of observations and measurements of physical facts, occurrences, and conditions, as they have occurred, excluding any material or information developed by means of computation or estimate. In the strictest sense, basic data include only the recorded notes of observations and measurements, although, in general use, it is taken to include computations or estimates necessary to present a clear statement of facts, occurrences, and conditions.

bearing housing   A protective outer shell placed around the bearing that contains lubricants and prevents contaminants from getting into the bearing.

bearing race   The grooved tracks that are fixed to the machine members in ball and roller bearing assemblies that contain the balls or rollers.

bell   The enlarged (female) end of a pipe into which an adjoining (male) pipe fits. Also, the flared pipe piece installed at the intake on a suction line that improves hydraulic conditions.

belt dressing   An aerosol product that increases operating efficiency and extends the life of drive belts. Belt dressing can reduce stiffness, restore flexibility, stop slipping, add power, reduce strain on bearings and bushings, and stop squealing.

biochemical oxygen demand (BOD)   A measure of the quantity of oxygen used in the biochemical oxidation of organic matter in a specified time, at a specific temperature, and under specified conditions.

biochemical oxygen demand (BOD) load   The BOD content, typically expressed in pounds per unit of time, of wastewater passing into a waste treatment system or to a body of water.

biofilm   Accumulation of microbial growth on the surface of a support material.

bleed   (1) To drain a liquid or gas, as to vent accumulated air from a water line or to drain a trap or a container of accumulated water. (2) The exuding, percolation, or seeping of a liquid through a surface.

blower   A single or multistage mechanical device, such as a compressor, that produces a flow of pressurized air.

boundary layer viscous drag   A force caused by the movement of a body through a fluid. The direction of this force is along the axis parallel to the movement. This force is used to create the pumping action in disc pumps.

bubbler level measurement   A fluid-measuring device that operates on the principle that a small constant flowrate of air bubbling into a liquid will produce a back pressure equal to the static head. Typically, the bubbler pressure is converted to a voltage or current signal that is fed into an electronic network. Variations in the signal level are used to turn pumps on or off, or adjust their speed.

Buna K   An acrylonitrile-butadiene rubber formulation with a higher acrylonitrile content than Buna N. This formula offers superior petroleum-based oil and fuel resistance than Buna N. Temperature and pressure performance are decreased, however. See also *Buna N*.

Buna N   A copolymer of butadiene and acrylonitrile that is also known as *NBR* (acrylonitrile-butadiene rubber) and also as *Nitrile* and as *Perbunan*. The acrylonitrile content of the copolymer varies from 18 to 50%. As the acrylonitrile content is increased, the resistance to oil and fuel of the polymer improves while the flexibility of the material at lower temperatures decreases. Buna N compounds have good resistance to alkaline and salt solutions, petroleum oils, vegetable oils, alcohols, glycols, gasolines, digester oils, and water. They are not suitable for use with strong oxidizing agents, chlorinated solvents, nitrated hydrocarbons, ketones, acetates, or aromatic hydrocarbons. The ozone and weather resistance of the materials is generally poor. The compound is used to make gaskets, hoses, and clothing. See also *Buna K*.

butt square joints   Joints in which the materials being joined butt against each other rather than overlap.

butterfly valve   A valve in which the disk, as it opens or closes, rotates about a spindle supported by the frame of the valve. The valve is opened at a stem. At full opening, the disk is in a position parallel to the axis of the conduit.

bypass   An arrangement of pipes, conduits, gates, and valves by which the flow may be passed around a hydraulic structure appurtenance or treatment process; a controlled diversion.

bypass line   Typically, a short segment of piping with isolation valves that allow flow to be diverted around a hydraulic structure, component, or treatment unit that may require periodic maintenance. The bypass line allows the main line to stay in service while the unit or component is worked on.

calcium hypochlorite   A solid that, when mixed with water, liberates the hypochlorite ion OCl⁻ and can be used for disinfection.

capacitance   The ability to store an electric charge, measured in farads. The capacitance of a capacitor is one farad when one coulomb of electricity changes the potential between the plates by 1 V.

capacity   (1) The quantity that can be contained exactly, or the rate of flow that can be carried out exactly. (2) The load for which a machine, apparatus, station, or system is rated.

carcinogen   A material that induces excessive or abnormal cellular growth in an organism.

carrying capacity   The maximum rate of flow that a conduit, channel, or other hydraulic structure is capable of passing.

cathodic protection   An electrical system for prevention of rust, corrosion, and pitting of steel and iron surfaces in contact with water. A low-voltage current is made to flow through the liquid or soil in contact with the metal in such a manner that the external electromotive force renders the metal structure cathodic and concentrates corrosion on auxiliary anodic parts used for that purpose.

cation   A positively charged ion attracted to the cathode under the influence of electrical potential.

cavitation   The action of an operating centrifugal pump, when it is attempting to discharge more water than suction can provide. Pockets of very low pressure form around the impeller. When this low pressure drops below the vapor pressure of the fluid being pumped, vapor bubbles form. As these bubbles move through the pump, the pressure increases, causing the bubbles to collapse with great force. Cavitation frequently causes pitting and gouging of the affected pump's impeller.

Celsius   The international name for the centigrade scale of temperature, on which the freezing point and boiling point of water are 0 °C and 100 °C, respectively, at a barometric pressure of 1.013×10^5 Pa (760 mm Hg).

Centigrade   A thermometer temperature scale in which 0° marks the freezing point and 100° the boiling point of water at 760 mm Hg barometric pressure. Also called Celsius. To convert temperature on this scale to Fahrenheit, multiply by 1.8 and add 32.

centrifugal non-clog pump   A type of centrifugal pump in which the impeller is rounded and free of sharp corners to minimize the chance that it will collect rags and stringy objects. The impeller is specially designed with clearances to pass larger solids.

centrifugal pump   A pump consisting of an impeller fixed on a rotating shaft and enclosed in a casing having an inlet and a discharge connection. The rotating impeller creates pressure in the liquid by the velocity derived from centrifugal force.

certification   A program to substantiate the capabilities of personnel by documentation of experience and learning in a defined area of endeavor.

cfs (cu ft/sec)   The rate of flow of a material in cubic feet per second; used for measurement of water, wastewater, or gas; equals 2.832×10^{-2} m^3/s.

chain and sprocket   A device used for the transmission of power where shafts are separated and the use of gears is impractical. Sprockets take the place of gears and drive one another by means of the chain passing over the sprocket teeth. A sprocket is a wheel with teeth shaped to mesh with the chain.

chamber   Any space enclosed by walls or a compartment; often prefixed by a descriptive word indicating its function, such as grit chamber, screen chamber, discharge chamber, or flushing chamber.

check valve   A valve designed to open in the direction of normal flow and close with reversal of flow. Check valves are typically of substantial construction, positive in closing, and permit no leakage in a direction opposite of normal flow. Two common types of check valves used in wastewater applications are ball check valves and flapper check valves.

chemical   Commonly, any substance used in or produced by a chemical process. Certain chemicals may be added to water or wastewater to improve treatment efficiency; others are pollutants that require removal.

chemical analysis   Analysis by chemical methods to show the composition and concentration of substances.

chemical conditioning of sludge   The addition of chemicals to a sludge before dewatering to improve the solids separation characteristics. Typical conditioners include polyelectrolytes, iron salts, and lime.

chemical dose   A specific quantity of chemical applied to a specific quantity of fluid for a specific purpose.

chemical feed pump   A pump that dispenses a chemical solution at a predetermined rate, for example, for the treatment of water, wastewater, or sludge.

chemical metering   The use of a variable speed pump and control system to adjust the dose of chemical being used for a particular application.

chemical reaction   A transformation of one or more chemical species into other species, resulting in the evolution of heat or gas, color formation, or precipitation. The reaction may be initiated by a physical process such as heating or by the addition of a chemical reagent, or it may occur spontaneously.

chemical reagent   A chemical added to a system to induce a chemical reaction.

chemical slurry   A thin mixture or suspension of fine particles of a chemical in a liquid, typically water.

chlorination The application of chlorine or chlorine compounds to water or wastewater, generally for the purpose of disinfection, but frequently for chemical oxidation and odor control.

chlorine (Cl₂) An element ordinarily existing as a greenish-yellow gas about 2.5 times heavier than air. At atmospheric pressure and a temperature of −30.1 °F (−48 °C), the gas becomes an amber liquid about 1.5 times heavier than water. Its atomic weight is 35.457, and its molecular weight is 70.914.

chlorine dose The amount of chlorine applied to a wastewater, typically expressed in milligrams per liter (mg/L) or pounds per million gallons (lb/mil. gal).

chopper pump A variation of the centrifugal pump that is better suited to pumping fluids containing solids and stringy materials. The edges of the impeller vanes are sharpened to allow them to cut or chop solid materials as they enter the pump, thus making it easier to pass them. Some variations also include a cutter bar that attaches to the end of the shaft to provide additional cutting ability.

CIP (capital improvement program) A plan that identifies and estimates the nature, schedule, cost, priority, and financing of long-term assets that an agency intends to build, rehabilitate, or acquire during a specific period.

circuit breaker Commonly called a "breaker", a device designed to open and close a circuit by non-automatic means and to open the circuit automatically on a predetermined overcurrent without damage to the breaker itself when properly applied within its rating. A breaker is a form of a switch, but a "switch" is not designed for the interruption of short-circuit currents, while a circuit breaker is.

clean-out hole A hole or port on the suction or discharge side of a pump impeller that is provided for inspection and maintenance purposes.

close-coupled pump A type of pump in which the pump shaft is integral with the motor shaft.

closed centrifugal pump A centrifugal pump having its impeller built with the vanes enclosed within circular disks.

closed conduit Any closed artificial or natural duct for conveying fluids.

closed impeller An impeller having the sidewalls extended from the outer circumference of the suction opening to the vane tips.

coating A material applied to the inside or outside of a pipe, valve, or other fixture to protect it primarily against corrosion. Coatings may be of various materials.

coliform-group bacteria A group of bacteria predominantly inhabiting the intestines of man or animals, but also occasionally found elsewhere. It includes all aerobic and facultative anaerobic, Gram-negative, nonspore-forming, rod-shaped bacteria that ferment lactose with the production of gas. Also included are all bacteria that produce a dark, purplish-green metallic sheen by the membrane filter technique used for coliform identification. The two groups are not always identified, but they are generally of equal sanitary significance.

collection system   The pipes, conduits, pumping stations, and appurtenances involved in the collection and transportation of wastewater and stormwater.

combined sewer   A sewer designed to receive both sanitary wastewater and stormwater.

combustible-gas indicator   An explosimeter; a device for measuring the concentration of potentially explosive fumes. The measurement is based on the catalytic oxidation of a combustible gas on a heated platinum filament that is part of a Wheatstone bridge.

comminution   An in-stream process of cutting and screening solids contained in wastewater flow.

comminutor   A shredding or grinding device that reduces the size of suspended materials in wastewater without removing them from the liquid.

compressed air   Air that has been reduced in volume, thereby exerting a pressure.

compression ring   A fitting used in certain kinds of pump seals to maintain pressure against the shaft packing.

concentric reducer   A pipe fitting that has flanges on both ends and that is used to transition between pipes with different sizes. These circular flanges share a common center.

confined space   A space that, by design, has limited openings for entry and exit; has unfavorable natural ventilation that could contain, collect, or produce dangerous air contaminants; and that is not intended for continuous human occupancy.

connecting rod   A plunger pump component that connects the shaft to the plunger.

continuous-flow pump   A displacement pump within which the direction of flow of the water is not changed or reversed.

contracted weir   A rectangular notched weir with a crest width narrower than the channel across which it is installed and with vertical sides extending above the upstream water level, producing a contraction in the stream of water as it leaves the notch.

contractual funding obligations   Incremental deposits made into various reserve accounts so that an agency may meet its minimum funding requirements.

conversion factor   A numerical constant by which a quantity with its value expressed in units of one kind is multiplied to express the value in units of another kind.

corrosion   The gradual deterioration or destruction of a substance or material by chemical action, frequently induced by electrochemical processes. The action proceeds inward from the surface.

corrosion control   (1) The sequestration of metallic ions and the formation of protective films on metal surfaces by chemical treatment. (2) Any action taken with the direct purpose being to prevent or slow the corrosion, erosion, or oxidation of a material. In regard to pumping systems, corrosion control might include controlling humidity, applying a proper paint system, providing cathodic protection, and so on. (3) In water treatment, any method that keeps the metallic ions of a conduit from going into solution, such as increasing the pH of the water, removing free oxygen from the water, or controlling the carbonate balance of the water.

coupling   A device for connecting two adjacent parts together. Couplings are commonly used in pumps to connect the pump shaft with the motor shaft.

crankshaft   A shaft for transmitting motion, consisting of a series of cranks and crankpins to which the connecting rods of an engine are attached.

cross-connection   (1) A physical connection through which a supply of potable water could be contaminated or polluted. (2) A connection between a supervised potable water supply and an unsupervised supply of unknown potability.

data   Records of observations and measurements of physical facts, occurrences, and conditions reduced to written, graphical, or tabular form.

debris   Generally solid wastes from natural and man-made sources deposited indiscriminately on land and water and often found in pipe inverts.

decomposition of wastewater   (1) The breakdown of organic matter in wastewater by bacterial action, either aerobic or anaerobic. (2) Chemical or biological transformation of the organic or inorganic materials contained in wastewater.

degree   (1) On the centigrade or Celsius thermometer scale, 1/100 of the interval from the freezing point to the boiling point of water under standard conditions; on the Fahrenheit scale, 1/180 of this interval. (2) A unit of angular measure; the central angle subtended by 1/360 of the circumference of a circle.

deposition   The act or process of settling solid material from a fluid suspension.

design criteria   (1) Engineering guidelines specifying construction details and materials. (2) Objectives, results, or limits that must be met by a facility, structure, or process in performance of its intended functions.

design flow   Engineering guidelines that typically specify the amount of influent flow that can be expected on a daily basis over the course of a year. Other design flows can be set for monthly or peak flows.

diaphragm   A dividing membrane or thin flexible partition.

diaphragm pump   A positive-displacement pump in which a flexible diaphragm, generally made of rubber or an equally resilient material, is the operating part. It is fastened at the edges in a vertical cylinder. When the diaphragm is raised, suction is exerted, and when it is depressed, the liquid is forced through a discharge valve.

diaphragm-type pressure gauge   The measurement from this type of gauge is taken from the movement of the diaphragm, located near the connection to the main line, because of pressure changes in the line. These types of gauges are very useful for measuring the pressure of sludge lines because they are not as prone to clogging as conventional gauges.

differential plunger pump   A reciprocating pump with a plunger so designed that it draws the liquid into the cylinder on the upward stroke, but is double acting on the discharge stroke.

direct current   An electric current that flows in one direction only and is substantially constant in value.

discharge   The flow or rate of flow from a canal, conduit, pump, stack, tank, or treatment process. See also *effluent*.

discharge area   The cross-sectional area of a waterway. Used to compute the discharge of a stream, pipe, conduit, or other carrying system.

discharge capacity   The maximum rate of flow that a conduit, channel, or other hydraulic structure is capable of passing.

discharge head   A measure of the pressure exerted by a fluid at the point of discharge, typically from a pump.

discharge pressure switch   A switch that is energized at preset pressures and that, when mounted on the discharge side of the pump, is used to provide feedback on the operation status of the pump.

discharge rate   (1) The determination of the quantity of water flowing per unit of time in a stream channel, conduit, or orifice at a given point by means of a current meter, rod float, weir, pitot tube, or other measuring device or method. The operation includes not only the measurement of velocity of water and the cross-sectional area of the stream of water, but also the necessary subsequent computations. (2) The numerical results of a measurement of discharge, expressed in appropriate units.

discharge valve   A valve located immediately downstream of a pump that can be used for throttling or isolating the discharge piping from the pump.

disinfectant   A substance used for disinfection and in which disinfection has been accomplished.

disinfection   (1) The killing of waterborne fecal and pathogenic bacteria and viruses in potable water supplies or wastewater effluents with a disinfectant; an operational term that must be defined within limits, such as achieving an effluent with no more than 200 colonies fecal coliform/100 mL. (2) The killing of the larger portion of microorganisms, excluding bacterial spores, in or on a substance, with the probability that all pathogenic forms are killed, inactivated, or otherwise rendered nonvirulent.

distributed control system (DCS)   An instrumentation/control system where a system of microprocessors located near the equipment they control are linked through a communications network. The overall status of the system can be monitored and operating parameters can be changed at a central operator control station that is also connected to the communications network.

diurnal   (1) Occurring during a 24-hour period; diurnal variation. (2) Occurring during the day (as opposed to night). (3) In tidal hydraulics, having a period or cycle of approximately 1 tidal day.

domestic wastewater   Wastewater derived principally from dwellings, institutions, and the like. It may or may not contain groundwater, surface water, or stormwater.

double-suction impeller   An impeller with two suction inlets, one on each side of the impeller.

double-suction pump   A centrifugal pump with suction pipes connected to the casing from both sides. These two suction inlets are located on either side of the pump's impeller.

dowel   A pin or stud, with rounded or beveled ends, fastened in one piece to locate it in proper position on a mating part.

drain   (1) A conduit or channel constructed to carry off, by gravity, liquids other than wastewater, including surplus underground, storm, or surface water. It may be an open ditch, lined or unlined, or a buried pipe. (2) In plumbing, any pipe that carries water or wastewater in a building drainage system.

drain valve   A valve located at a low point on a pump or other device or piece of equipment that can be opened when the device is taken out of service to allow it to drain by gravity.

drawdown   (1) The magnitude of the change in surface elevation of a body of water as a result of the withdrawal of water. (2) The magnitude of the lowering of the water surface in a well, and of the water table or piezometric surface adjacent to the well, resulting from the withdrawal of water from the well by pumping. (3) In a continuous water surface with accelerating flow, the difference in elevation between downstream and upstream points.

drip-proof motor   A type of squirrel cage motor designed to be open to the atmosphere for cooling the windings. The ventilation openings on these motors are constructed so that water falling on the motor from a vertical angle of no greater than 15 deg will not enter the motor. These motors are used for indoor or weatherproof and dust-free applications.

dry weather flow   (1) The flow of wastewater in a combined sewer during dry weather. Such flow consists mainly of wastewater, with no stormwater included. (2) The flow of water in a stream during dry weather, typically contributed to entirely by groundwater.

dry well   A dry compartment in a pumping station, near or below pumping level, where the pumps are located.

dynamic head   (1) When there is flow, (a) the head at the top of a waterwheel; (b) the height of the hydraulic grade line above the top of a waterwheel; (c) the head against which a pump works. (2) That head of fluid that would produce statically the pressure of a moving fluid.

dynamic suction head   The reading of a gauge on the suction line of a pump corrected for the distance of the pump below the free surface of the body of liquid being pumped; exists only when the pump is below the free surface. When pumping proceeds at the required capacity, the vertical distance from the source of supply to the center of the pump minus velocity head and entrance and friction losses. Internal pump losses are not subtracted.

dynamic suction lift   When pumping proceeds at the required capacity, the vertical distance from the source of supply to the center of the suction end of a pump, plus velocity head and entrance and friction losses. Internal pump losses are not added.

E. coli   *See Escherichia coli.*

eccentric   A disk or wheel whose axis is offset from its true center so that it is capable of imparting a reciprocating motion when it revolves.

eccentric reducer   A pipe fitting with two different-sized flanges at each end that is used to join two pipes of different sizes. These circular flanges are offset eccentrically (i.e., they share a common invert).

eductor   A device used to mix air with water; liquid pump operating under a jet principle, using liquid under pressure as the operating medium to enhance air in the liquid.

efficiency   The relative results obtained in any operation in relation to the energy or effort required to achieve such results. It is the ratio of the total output to the total input, expressed as a percentage.

effluent   Wastewater or other liquid, partially or completely treated or in its natural state, flowing out of a reservoir, basin, water resource recovery facility, or industrial treatment facility, or part thereof. See also *influent*.

elbow   A pipe fitting that connects two pipes at an angle. The angle is always 90 deg, unless another angle is stated.

elevation head   The energy possessed per unit weight of a fluid because of its elevation above some point. Also called position head or potential head.

emergency lighting   Battery-operated lighting units to provide illumination in the event of a power failure. Their main purpose is to improve safety in emergency situations.

end suction pump   A type of horizontal centrifugal pump with its intake located on the side parallel to the pump shaft.

EPDM   Ethylene-propylene-diene monomer, or EPDM, is an elastomer that is generally blended or copolymerized with other elastomers and modifiers to substantially improve the performance of the final EPDM elastomer product. Originally developed for the automobile tire industry, EPDM is very resistant to water logging as well as most water-based chemicals. It also has a very inert structure that remains stable over long periods of time.

Escherichia coli (E. coli)   One of the species of bacteria in the fecal coliform group. It is found in large numbers in the gastrointestinal tract and feces of warm-blooded animals and man. Its presence is considered indicative of fresh fecal contamination, and it is used as an indicator organism for the presence of less easily detected pathogenic bacteria.

explosimeter   A device for measuring the concentration of potentially explosive fumes. Also called a *combustible-gas indicator*.

explosive gases or vapors   Ignitable or flammable gases or vapors (e.g., acetylene, hydrogen, gasoline, ethylene).

facultative   Having the ability to live under different conditions, for example, with or without free oxygen.

facultative bacteria   Bacteria that can grow and metabolize in the presence and absence of dissolved oxygen.

Fahrenheit   A temperature scale in which 32° marks the freezing point and 212° the boiling point of water at 760 mm Hg. To convert to centigrade (Celsius), subtract 32 and multiply by 0.5556.

flange   A projecting rim, edge, lip, or rib.

flanged coupling   A type of rigid coupling that consists of two flanged rigid members with several bolt holes for connecting the two pieces.

flap-type check valve   A valve whose flap is hinged at one end and, therefore, opens and shuts by rotating about the hinge.

flexible coupling   A type of coupling that transmits driving torque as well as accommodates minor misalignments between pump and motor shafts.

float control   A float device that is triggered by changing liquid levels and that activates, deactivates, or alternates process equipment operation.

float gauge   A device for measuring the elevation of the liquid, the actuating element of which is a buoyant float that rests on the surface of the liquid and rises or falls with it. The elevation of the surface is measured by a chain or tape attached to the float.

float switch   An electrical switch operated by a float in a tank or reservoir and typically controlling the motor of a pump.

float valve   A valve, such as a plug or gate, that is actuated by a float to control the flow into a tank.

floating bushing   A bushing that can move radially (and thereby center itself around the pump shaft) that is used in some mechanical pump seals.

flooded suction   (1) The head at the inlet to a pump. (2) The head below atmospheric pressure in a piping system. A suction head exists when liquid is pumped from an open-top tank and the liquid level is above the centerline of the pump. See also *suction head.*

flow   (1) The movement of a stream of water or other fluid from place to place; the movement of silt, water, sand, or other material. (2) The fluid that is in motion. (3) The quantity or rate of movement of a fluid discharge; the total quantity carried by a stream. (4) To issue forth or discharge. (5) The liquid or amount of liquid per unit time passing a given point.

flow meter   A device that indicates the flowrate of the pump and verifies that the pump is pumping fluid.

flow-control valve   A device that controls the rate of flow of a fluid.

flowrate   The volume or mass of a gas, liquid, or solid material that passes through a cross section of conduit in a given time; measured in such units as kilograms per hour (kg/h), cubic meters per second (m³/s), liters per day (L/d), or gallons per day (gpd).

flow recording   Documentation of the rate of flow of a fluid past a given point. The recording is typically accomplished automatically.

flow sensor   A device that measures (senses) a parameter, such as depth, velocity, or pressure drop, that is related to flowrate. Flowrates in various water resource recovery facility streams are measured to make process control decisions.

flushing   The flow of water under pressure in a conduit or well to remove clogged material.

foot valve   (1) A valve placed at the bottom of the suction pipe of a pump that opens to allow water to enter the suction pipe, but closes to prevent water from passing out of it at the bottom end. (2) A valve with the reverse action attached to the drainage pipe of a vacuum chamber. It allows water to drain out, but closes to hold the vacuum.

force main   A pressurized pipeline that carries pumped water or wastewater.

friction factor   A measure of the resistance to flow of fluid in a conduit as influenced by wall roughness.

friction head   The head lost by water flowing in a stream or conduit as the result of the disturbances set up by the contact between the moving water and its containing conduit and by intermolecular friction. In laminar flow, the head lost is approximately proportional to the first power of the velocity; in turbulent flow to a higher power, approximately the square of the velocity. Although, strictly speaking, head losses such as those caused by bends, expansions, obstructions, and impact are not included in this term, the typical practice is to include all such head losses under this term.

friction loss   The head lost by water flowing in a stream or conduit as the result of the disturbances set up by the contact between the moving water and its containing conduit and by intermolecular friction. See also *friction head*.

fuse   A protective device that carries the full current of a circuit. If the current is higher than the fuse rating, it contains a substance that will melt and break the current. Fuses cannot be reset, but must be replaced.

gate valve   A type of valve used for isolation or throttling in which the closing element (the gate) is a disk that moves across the flow in a groove or slot to provide support against pressure.

gauge   (1) A device for indicating the magnitude or position of an element in specific units when such magnitude or position is subject to change; examples of such elements are the elevation of a water surface, the velocity of flowing water, the pressure of water, the amount or intensity of precipitation, and the depth of snowfall. (2) The act or operation of registering or measuring the magnitude or position of a thing when these characteristics are subject to change. (3) The operation of determining the discharge in a waterway by using both discharge measurements and a record of stage.

gland nut   A fitting used to adjust the pressure on the packing to control seal water leakage.

globe valve   A valve having a round, ball-like shell and horizontal disk.

gpd   The rate of water, wastewater, or other flow measured in gallons per day.

gpdc   The rate of water, wastewater, or other flow measured in gallons per capita of served population per day.

gpm   The rate of water, wastewater, or other flow measured in gallons per minute.

gradient   The rate of change of any characteristic per unit of length or slope. The term is typically applied to such things as elevation, velocity, or pressure. See also *slope*.

grease and oil   In wastewater, a group of substances including fats, waxes, free fatty acids, calcium and magnesium soaps, mineral oils, and certain other nonfatty materials;

water-insoluble organic compounds of plant and animal origins or industrial wastes that can be removed by natural flotation skimming.

grease relief fitting   A port on a greased chamber that is located so that when new grease is added, the buildup of pressure pushes the used grease out through this port.

grinder pump   A mechanical device that shreds solids and raises wastewater to a higher elevation through pressure sewers. These pumps are used to handle raw domestic or industrial wastewater, sludge, septage, and to break up digester scum.

grit   Small, dense mineral particles such as sand, coffee grounds, and gravel that are suspended in wastewater. Grit causes excessive wear and tear on pumps and other equipment in collection systems. It is removed in a grit chamber at the water resource recovery facility to minimize abrasion and the wearing of subsequent treatment devices.

grit chamber   A detention chamber or an enlargement of a sewer designed to reduce the velocity of flow of the liquid to permit the separation of mineral solids (grit) from organic solids by differential sedimentation.

grit collector   A device placed in a grit chamber to convey deposited grit to a point of collection.

grit separator   Any process or device designed to separate grit from a water or wastewater stream.

ground   A circuit that has the same potential as the earth within the facility. It can act as one leg of the electrical circuit.

hazardous waste   Waste that is potentially damaging to environmental health because of toxicity, ignitability, corrosivity, chemical reactivity, or other reasons.

head   (1) The height of the free surface of fluid above any point in a hydraulic system; a measure of the pressure or force exerted by the fluid. (2) The energy, either kinetic or potential, possessed by each unit weight of a liquid, expressed as the vertical height through which a unit weight would have to fall to release the average energy possessed. It is used in various compound terms such as pressure head, velocity head, and loss of head. (3) The upper end of anything, such as headworks. (4) The source of anything, such as a headwater. (5) A comparatively high promontory with either a cliff or steep face extending into a large body of water, such as a sea or lake. An unnamed head is typically called a headland.

head gate   A gate at the entrance to a conduit such as a pipeline, penstock, or canal.

head loss   Energy losses resulting from the resistance of flow of fluids; may be classified into conduit surface and conduit form losses.

header   (1) A structure installed at the head or upper end of a gully to prevent overfall cutting. (2) A supply ditch for the irrigation of a field. (3) A large pipe installed to intercept the ends of a series of pipes; a manifold. (4) The closing plate on the end of a sewer lateral that will not be used immediately.

headworks   (1) All the structures and devices located at the head or diversion point of a conduit or canal. The term, as used, is practically synonymous with diversion works;

an intake heading. (2) The initial structures and devices of a water or water resource recovery facility.

heat sensor   A sensor, typically placed on bearings or in motor windings, that activates an alarm and/or shuts down a unit when a preset high temperature is reached.

horsepower (hp)   Common unit of power (i.e., rate at which work is done). One horsepower is the power necessary to lift 33 000 lb of mass 1 ft in 1 min.

hour meter   A device to log the operational use of a pump or other equipment in hours.

hydraulic radius   The cross-sectional area of a stream of water divided by the length of that part of its periphery in contact with its containing conduit; the ratio of area to wetted perimeter. Also called hydraulic mean depth.

hydrogen ion concentration   The concentration of hydrogen ions in moles per liter of solution (moles/L). Commonly expressed as the pH value, which is the logarithm of the reciprocal of the hydrogen ion activity. See also *pH*.

hydrogen sulfide (H_2S)   A toxic and lethal gas produced in sewers and digesters by anaerobic decomposition. Detectable in low concentrations by its characteristic "rotten egg" odor, it deadens the sense of smell in higher concentrations or after prolonged exposure. Respiratory paralysis and death may occur quickly at concentrations as low as 0.07%, by volume, in air.

Hypalon   The trade name for a chlorosulfonated polyethylene, or CSPE, -based elastomer that exhibits high resistance to weathering, ozone, acids/bases, alcohol, heat, oil, flame, and abrasion.

idler pulley   A pulley for a belt or chain that serves to guide the belt or to take up slack (also known as an "idler").

impeller   A rotating set of vanes designed to impel rotation of a mass of fluid.

indicator gauge   A gauge that shows, by means of an index, pointer, or dial, the instantaneous value of such characteristics as depth, pressure, velocity, stage, and discharge or the movements or positions of waste-controlling devices. See also *recorder*.

induction motor   An alternating-current motor in which torque is produced by the reaction between a varying magnetic field generated in the stator and the current induced in the coils of the rotor.

industrial wastewater   Wastewater derived from industrial sources or processes.

infectious hepatitis   An acute viral inflammation of the liver characterized by jaundice, fever, nausea, vomiting, and abdominal discomfort; may be waterborne.

infiltration   (1) The flow or movement of water through the interstices or pores of a soil or other porous medium. (2) The quantity of groundwater that leaks into a pipe through joints, porous walls, or breaks. (3) The entrance of water from the ground into a gallery. (4) The absorption of liquid by the soil, either as it falls as precipitation or from a stream flowing over the surface.

infiltration/inflow   Infiltration is groundwater that leaks into the sewerage system through pipe joints and defects. *Inflow* refers to water that enters sewers from improperly connected catch basins, sump pumps, and land and basement drains and defective manholes. Inflow also enters through improperly closed or defective harbor combined sewer overflow tidegates when the tide is high.

inflow   In relation to sanitary sewers, the extraneous flow that enters a sanitary sewer from sources other than infiltration, such as roof leaders, basement drains, land drains, and manhole covers. See also *infiltration/inflow*.

influent   Water, wastewater, or other liquid flowing into a reservoir, basin, water resource recovery facility, or treatment process. See also *effluent*.

inlet   (1) A surface connection to a drain pipe. (2) A structure at the diversion end of a conduit. (3) The upstream end of any structure through which water may flow. (4) A form of connection between the surface of the ground and a drain or sewer for the admission of surface or stormwater. (5) An intake.

inorganic   All those combinations of elements that do not include organic carbon. See also *organic*.

inorganic matter   Mineral-type compounds that are generally nonvolatile, not combustible, and not biodegradable. Most inorganic-type compounds or reactions are ionic in nature; therefore, rapid reactions are characteristic.

in-rush current   A high current draw that occurs during startup of a piece of equipment with a standard full-voltage starter.

instrumentation   Use of technology to control or monitor pumps, motors, or levels within pumping stations and headworks.

isolation valve   A valve that is placed before or after a piece of equipment that may need to be removed from service.

journal   The part of a shaft or axle that turns in or against a bearing.

laminar flow   The flow of a viscous fluid in which particles of the fluid move in parallel layers, each of which has a constant velocity but is in motion relative to its neighboring layers. Also called streamline flow, viscous flow.

level sensor   A device that detects changes in liquid level, such as a float in a tank.

lift station   A structure that contains pumps and appurtenant piping, valves, and other mechanical and electrical equipment for pumping water, wastewater, or other liquid. Also called a pumping station.

liquid level sensor   A device, such as a bubbler system or ultrasonic level sensor, for monitoring the level of fluid in a wet well.

local start/stop button station   A simple control station located adjacent to the piece of equipment it controls to allow operators to start or stop the equipment. These local stations are important safety features in case of an emergency.

long radius bend   A pipe bend that has a more gradual turning radius than standard bends, designed to reduce head loss.

Manning formula   A formula for open-channel flow published by Manning in 1890. See also *Manning roughness coefficient*.

Manning roughness coefficient The roughness coefficient in the Manning formula for determination of the discharge coefficient.

manometer An instrument for measuring pressure. It typically consists of a U-shaped tube containing a liquid, the surface of which moves proportionally in one end of the tube with changes in pressure in the liquid in the other end; also, a tube type of differential pressure gauge.

manual pump control Push-button stations or selector switches that energize or de-energize the pump motor starter. Push-button stations (three-wire control) are electrically interlocked so that units have to be restarted manually after a power outage, whereas a selector switch (two-wire control) remains in the on position and restarts automatically.

mechanical seal A component of centrifugal pumps that prevents leakage along the shaft. Mechanical seals may be composed of carbon or stainless steel, ceramics, tungsten carbide, or brass. Water or oil can be used for lubrication of moving parts. They substitute for packing. Mechanical seals may be internal or external to stuffing boxes.

meter (1) An instrument for measuring some quantity such as the rate of flow of liquids, gases, or electric currents. (2) A unit of length equivalent to 1000 mm or 3.281 ft.

methane (CH_4) A colorless, odorless, flammable, gaseous hydrocarbon present in natural gas and formed by the anaerobic decomposition of organic matter, or produced artificially by heating carbon monoxide and hydrogen over a nickel catalyst. See also *anaerobic digestion*.

methane bacteria A specialized group of obligate anaerobic bacteria that decompose organic matter to form methane.

methane fermentation A reaction sequence that produces methane during the anaerobic decomposition or organic waste. In the first phase, acid-forming bacteria produce acetic acid; in the second, the methane bacteria use this acid and carbon dioxide to produce methane. Fermentation results in the conversion of organic matter into methane gas.

mg/L Milligrams per liter; a measure of concentration equal to and replacing parts per million (ppm) in the case of dilute concentrations.

mgd Million gallons per day; a measure of flow equal to 1.547 cfs (ft^3/sec), 681 gpm, or 3.785 m^3/d.

microorganisms Very small organisms, either plant or animal, invisible or barely visible to the naked eye. Examples are algae, bacteria, fungi, protozoa, and viruses.

minor loss The component of the total head of a pumping system attributable to the change in velocity and/or direction of the fluid as it moves through the pipe fittings.

mixed-flow impeller A type of impeller that develops its head mostly by centrifugal force and partly by axial thrust imparted to the wastewater. It is used for relatively high flow capacities against relatively low heads. These impellers are suitable for handling wastewater and stormwater.

monitoring (1) Routine observation, sampling, and testing of designated locations or parameters to determine the efficiency of treatment or compliance with standards or requirements. (2) The procedure or operation of locating and measuring radioactive contamination by means of survey instruments that can detect and measure, as dose rate, ionizing radiations.

motor   A device that converts (typically electrical) energy into mechanical power.

motor controller   A specialized type of controller, the typical functions of which include starting, accelerating, stopping, reversing, and protecting motors.

motor efficiency   The ratio of mechanical power output of a motor to the electrical power input to the motor.

motor enclosure   A protective casing installed around a motor.

motor starter   A piece of control equipment designed to make and break the electrical power connection to a motor as well as to provide overload protection.

multi-loop controller   A self-contained unit that can combine the function of several single-loop controllers. These devices are ideal for systems that have several different functions that all need to be simultaneously controlled.

multistage pump   A centrifugal pump with two or more sets of vanes or impellers connected in series in the same casing. Such a pump may be designated as two-stage, three-stage, or more, according to the number of sets of vanes used. The purpose is to increase the head of the discharging fluid.

National Pollutant Discharge Elimination System (NPDES)   A permit that sets the requirements for efficient discharge and that is the basis for the monthly monitoring reports required by most states in the United States.

negative head   (1) The loss of head in excess of the static head (a partial vacuum). (2) A condition of negative pressure produced by clogging of rapid sand filters near the end of a filter run.

negative pressure   A pressure less than the local atmospheric pressure at a given point.

negative suction head   A negative suction head or suction lift exists when liquid is pumped from an open-top tank and the liquid level is below the centerline of the pump. The magnitude of the suction lift is equal to the vertical distance from the liquid surface in the tank to the centerline of the pump. See also *suction lift*.

neoprene   An elastomer, also known as polychloroprene, with good resistance to weather, petroleum-based fluids, water, heat, and flame. It was originally developed as an oil-resistant substitute for natural rubber. The polymer structure can be modified by co-polymerizing chloroprene with sulfur and/or 2,3-dichloro-1,3-butadiene to yield a family of materials with a broad range of chemical and physical properties.

nitrogen (N)   An essential nutrient that is often present in wastewater as ammonia, nitrate, nitrite, and organic nitrogen. The concentrations of each form and the sum (total nitrogen) are expressed as milligrams per liter (mg/L) elemental nitrogen. Also present in some groundwater as nitrate and in some polluted groundwater in other forms.

nonclogging impeller   An impeller of the open, closed, or semi-closed type designed for passing large solids.

non-ventilated motor   A type of motor whose enclosure does not include any louvered ventilation openings.

odor control   Prevention or reduction of objectionable odors by chlorination, aeration, or other processes or by masking with chemical agents.

oiler   A mechanical device for furnishing a steady supply of oil to bearings or other parts.

open centrifugal pump   A centrifugal pump in which the impeller is built with a set of independent vanes.

open channel   Any natural or artificial water conduit in which water flows with a free surface.

open-channel flow   Flow of a fluid with its surface exposed to the atmosphere. The conduit may be an open channel or a closed conduit flowing partly full.

open impeller   An impeller without attached side walls.

open motor   A type of motor designed to be open to the atmosphere to cool the windings.

operators   (1) Persons employed to operate a treatment facility. (2) Mechanism used to manipulate valve positions.

organic   Volatile, combustible, and sometimes biodegradable chemical compounds containing carbon atoms (carbonaceous) bonded together with other elements. The principal groups of organic substances found in wastewater are proteins, carbohydrates, and fats and oils. See also *inorganic*.

orifice   (1) An opening with a closed perimeter, typically of regular form, in a plate, wall, or partition through which water may flow; generally used for the purpose of measurement or control of such water. The edge may be sharp or of another configuration. (2) The end of a small tube such as a pitot tube or piezometer.

o-ring seal   A ring-shaped gasket, typically made of synthetic rubber, used to make a joint fluid-tight.

outfall   (1) The point, location, or structure where wastewater or drainage discharges from a sewer, drain, or other conduit. (2) The conduit leading to the ultimate disposal area.

overcurrent relay   A relay that is preset to open at a certain amperage to shut down the equipment on that circuit. The preset amperage is selected above normal operating amperage of the equipment, but below that at which damage to the controls or the motor is likely to occur.

oxygen (O)   A chemical element necessary for many biological activities. It typically occurs as O_2 and constitutes approximately 20% of the atmosphere.

oxygen deficiency   (1) The additional quantity of oxygen required to satisfy the oxygen requirement in a given liquid; typically expressed in milligrams per liter (mg/L). (2) Lack of oxygen.

packing   Material used to minimize leakage around the pump shaft where it penetrates the volute casing. A number of different kinds of packing material are available, depending on the application.

parallel misalignment   A type of shaft misalignment that results from the pump shaft and the motor shaft not meeting cleanly at the coupling because the pump shaft's alignment and the motor shaft's alignment are offset and, therefore, run parallel to one another.

Parshall flume   A calibrated device developed by Parshall for measuring the flow of liquid in an open conduit. It consists essentially of a contracting length, a throat, and an expanding length. At the throat is a sill over which the flow passes. The upper and lower heads are each measured at a specific distance from the sill. The lower head need not be measured unless the sill is submerged more than about 67%.

partial vacuum   Description of a space condition in which the pressure is less than atmospheric.

parts per million (ppm)   The number of weight or volume units of a minor constituent present with each 1 million units of a solution or mixture. The more specific term, *milligrams per liter (mg/L)*, is preferred.

pathogenic bacteria   Bacteria that cause disease in the host organism by their parasitic growth.

pathogens   Pathogenic or disease-producing organisms.

peristaltic pump   A pump in which fluid is forced through flexible tubing or hose by waves of contraction produced mechanically by compression of the flexible conduit.

pH   A measure of the hydrogen-ion activity in a solution, expressed as the logarithm (base 10) of the reciprocal of the hydrogen-ion activity in gram moles per liter (g/mol/L). On the pH scale (0 to 14), a value of 7 at 25 °C (77 °F) represents a neutral condition. Decreasing values indicate increasing hydrogen-ion activity (acidity); increasing values indicate decreasing hydrogen-ion activity (alkalinity).

pipe   A closed conduit that diverts or conducts water or wastewater from one location to another.

pipe diameter   The nominal or commercially designated inside diameter of a pipe, unless otherwise stated.

pipe fittings   Connections, appliances, and adjuncts designed to be used in connection with pipes; examples are elbows and bends to alter the direction of a pipe; tees and crosses to connect a branch with a main; plugs and caps to close an end; and bushings, diminishers, or reducing sockets to couple two pipes of different dimensions.

pipe gallery   (1) Any conduit for pipe, typically of a size to allow a person to walk through. (2) A gallery provided in a water resource recovery facility for the installation of conduits and valves and used as a passageway to provide access to them.

piping system   A system of pipes, fittings, and appurtenances within which a fluid flows.

piston pump   A reciprocating pump in which the cylinder is tightly fitted with a reciprocating piston.

plumbing   (1) The pipes, fixtures, and other apparatus inside a building for bringing in the water supply and removing the liquid and waterborne wastes. (2) The installation of the foregoing pipes, fixtures, and other apparatus.

plumbing fixture   Receptacles that receive liquid, water, or wastewater and discharge them into a drainage system.

plunger crosshead   A component of the power end of a plunger pump that transfers the plunger load to the connecting rod.

plunger pump   A positive displacement pump that consists of a cylinder in which a plunger moves back and forth. The plunger does not come into contact with the cylinder walls, but enters and withdraws through packing glands. The packing may be inside or outside the center, depending on the pump design. As the plunger moves outward, the volume available in the cylinder increases, and fluid enters through a one-way inlet valve. As the plunger moves inward, the volume available in the cylinder decreases, the pressure of the fluid increases, and the fluid is forced out through an outlet valve.

pneumatic ejector   A device for raising wastewater, sludge, or other liquid by alternately admitting it through an inward swinging check valve into the bottom of an airtight compartment and then discharging it through an outward swinging check valve by letting compressed air into the compartment above the liquid.

pollution   (1) Specific impairment of water quality by agricultural, domestic, or industrial wastes (including thermal and atomic wastes) to a degree that has an adverse effect on any beneficial use of the water. (2) The addition to a natural body of water of any material that diminishes the optimal economic use of a waterbody by the population it serves and has an adverse effect on the surrounding environment.

polychlorinated biphenyls (PCBs)   A class of aromatic organic compounds with two six-carbon unsaturated rings, with chlorine atoms substituted on each ring and more than two such chlorine atoms per molecule of PCB. They are typically stable, resist both chemical and biological degradation, and are toxic to many biological species.

polyvinyl chloride (PVC)   An artificial polymer made from vinyl chloride monomer frequently used in pipes, sheets, and vessels for transport, containment, and treatment in water and wastewater facilities.

portable generator   Small skid-mounted emergency power supply that is easily moved to a location when needed. Portable generators are often used to power pumps in pumping stations to keep the station from flooding or backing up the influent sewers when the normal power supply is temporarily lost.

positive-displacement pump   A pump that lifts a given volume for each cycle of operation. Water is brought into the pump chamber by a vacuum created by the withdrawal of a piston or piston-like device, which, upon its return, displaces a certain volume of water contained in the chamber and forces it to flow through the discharge valve and pipe. Positive-displacement pumps can be divided into two main classes: reciprocating and rotary. Reciprocating pumps include piston, plunger, and diaphragm types; rotary pumps include lobe, screw, and peristaltic pumps. Positive-displacement pumps are typically used for pumping raw sludge and slurries and other high-density or viscous fluids. They may also be used for chemical application.

potable water   Water without contamination or pollution that is considered safe for human consumption.

pressure   (1) The total load or force acting on a surface. (2) In hydraulics, unless otherwise stated, the pressure per unit area or intensity of pressure above local atmospheric pressure, expressed in pounds per square inch (psi) or kilograms per square centimeter ($kg\ cm^2$).

pressure gauge   A device for measuring the pressure of liquids, gases, or solids, typically in a discharge line for monitoring pressure at the pump.

pressure-relief device   A device such as a valve that opens automatically to relieve stress on a pipeline when pressure reaches an assigned limit.

pressure-relief valve   Valve that opens automatically when the pressure reaches a preset limit to relieve stress on a pipeline.

pressure sensor   A standard switch in a hydropneumatic tank, used to sense pressure drop and to start pumps in water and wastewater systems or that is tied into an automatic control system to regulate a process.

programmable logic controller (PLC)   Devices that take signals from sensors on process inputs and outputs and through a logic-based program and produce a change in one or more manipulated variables by means of some type of actuator. Although PLCs are designed and programmed in mathematical terms, they implement a control strategy through mechanical or electronic means.

progressing cavity pump   A type of positive-displacement pump that is composed of a single-threaded rotor that operates with a minimum of clearance in a double-threaded helix, called a *stator*, that is typically made of rubber. These pumps are commonly used for sludge transfer.

propeller pump   A centrifugal pump that develops most of its head by the propelling or lifting action of the vanes on the liquid. Also called an axial-flow pump.

propeller-type impeller   An impeller of the straight axial-flow type.

proportional pump   A type of diaphragm pump whose flowrate is quite precisely adjustable by changing the displacement per stroke (by altering the stroke length) or varying the stroke speed. This type of pump is commonly used for chemical feed.

proportional weir   A special type of weir in which the discharge through the weir is directly proportional to the head.

pump   A mechanical device for causing flow, for raising or lifting water or other fluid, or for applying pressure to (compressing) fluids. Pumps are classified according to the way in which energy is imparted to the fluid. The basic methods are volumetric displacement, addition of kinetic energy, and use of electromagnetic force.

pump curve   A curve or curves showing the interrelation of speed, dynamic head, capacity, brake horsepower, and efficiency of a pump.

pump efficiency   The ratio of energy converted into useful work to the energy applied to the pump shaft, or the energy difference in the water at the discharge and suction nozzles divided by the power input at the pump shaft.

pump efficiency curve   A graphical representation of the operating characteristics of a pump. The curve shows the interrelation of flow and total dynamic head, the efficiency, the net positive suction head required, and the brake horsepower required. See also *pump performance curve*.

pump performance curve   A graphical representation of the operating characteristics of a pump. The curve shows the interrelation of flow and total dynamic head, the efficiency, the net positive suction head required, and the brake horsepower required.

pump performance test   A procedure, which can be conducted in a manufacturer's shop or at the installation location, that demonstrates whether the pump can meet its performance requirements for head, capacity, power, efficiency, and suction.

pump pit   A dry well or chamber, below ground level, in which a pump is located.

pump priming   Filling a pump with a liquid to be transmitted so that all air contained in the pump has been allowed to escape.

pump rotation   The direction in which a centrifugal pump impeller is turning. By design, impellers must have the correct rotation to push fluid out from the impeller.

pump sequencing   A maintenance management technique by which redundant pieces of equipment are alternately operated for unequal amounts of time. This allows for the periodic running of backup equipment to prevent seizing while not running all equipment for equal times to stagger maintenance requirements.

pump stage   The number of impellers in a centrifugal pump; for example, a single-stage pump has one impeller; a two-stage pump has two impellers.

pump stroke   The lineal distance traveled by the piston or plunger of a reciprocating pump through one-half of its cycle of movement.

pumping head   The sum of the static head and friction head on a pump discharging a given quantity of water.

pumping station   (1) A facility housing relatively large pumps and their appurtenances. Pump house is the typical term for shelters for small water pumps. (2) A facility containing lift pumps to facilitate wastewater collection or potable water distribution.

radial flow impeller   A turbulent mixer that causes tank fluid to flow perpendicular to the impeller's axis of rotation. Types of radial flow impellers include disk-style flat-blade turbines and curved-blade turbines. They are used in applications where high shear rates are needed, such as in dispersion processes.

radially split pump   A type of centrifugal pump that has a split in its casing that is perpendicular to its drive shaft.

raw wastewater   Wastewater before it receives any treatment.

receiving water   A river, lake, ocean, or other watercourse into which wastewater or treated effluent is discharged.

recessed impeller centrifugal pump   A type of centrifugal pump that features concentric casing in which the impeller is completely recessed, out of the volute area. The rotation of the recessed impeller causes a vortex action in the fluid being pumped. This type of pump is also known as a *vortex* or *torque flow pump*.

reciprocating pump   A type of displacement pump consisting essentially of a closed cylinder containing a piston or plunger as the displacing mechanism. Liquid is drawn into the cylinder through an inlet valve and forced out through an outlet valve. When the piston acts on the liquid in one end of the cylinder, the pump is termed single-action; when it acts in both ends, it is termed double-action.

recorder   (1) A device that makes a graph or other record of the stage, pressure, depth, velocity, or the movement or position of water-controlling devices, typically as a function of time. (2) The person who records the observational data.

recording gauge   An automatic instrument for measuring and recording graphically and continuously. Also called a register.

rectangular weir   A weir having a notch that is rectangular in shape.

relief valve   A valve that releases air from a pipeline automatically without loss of water, or introduces air to a line automatically if the internal pressure becomes less than that of the atmosphere.

repair   An element of maintenance, as distinguished from replacement or retirement.

replacement   Installation of new or alternate equipment in place of existing equipment for a variety of reasons, such as obsolescence, total disrepair, improvement, or modification.

replacement cost   (1) The actual or estimated cost of duplication with a property of equal utility and desirability. (2) The cost of replacing property.

return activated sludge (RAS)   Biological solids collected from the clarifier and recycled to the aeration tank.

rigid coupling   Inflexible coupling used to link a pump shaft to a drive shaft.

roller bearing   A bearing used to reduce friction in which the shaft or journal rotates in peripheral contact with a number of (typically steel) rollers usually contained in a cage.

rotary pump   A type of displacement pump consisting essentially of elements rotating in a pump case that is closely fit. The rotation of these elements alternately draws in and discharges the water being pumped. Such pumps act with neither suction nor discharge valves, operate at almost any speed, and do not depend on centrifugal forces to lift the fluid.

rotary valve   A valve consisting of a casing more or less spherical in shape and a gate that turns on trunnions through a 90-deg opening or closing and having a cylindrical opening of the same diameter as that of the pipe it serves.

rotor   A rotating part of a motor, dynamo, or electrical machine.

roughness coefficient   A factor, in many engineering equations, for computing the average velocity of flow of water in a conduit or channel. It represents the effect of the roughness of the confining material on the energy losses in the flowing water.

safety guards   Guards around exposed shafts and other moving parts to protect operators from accidental contact and from whipping if a shaft breaks.

safety valve   A valve that automatically opens when prescribed conditions, typically pressure, are exceeded in a pipeline or other closed receptacle containing liquids or gases. It prevents such conditions from being exceeded and causing damage.

sanitary sewer   A sewer that carries liquid and waterborne wastes from residences, commercial buildings, industrial facilities, and institutions together with minor quantities of ground, storm, and surface water that are not admitted intentionally. See also *wastewater*.

screen   A device with openings, generally of uniform size, used to retain or remove suspended or floating solids in a flow stream, preventing them from passing a given point in a conduit. The screening element may consist of parallel bars, rods, wires, grating, wire mesh, or perforated plate.

screening   A preliminary treatment process that removes large suspended or floating solids from raw wastewater to prevent subsequent plugging of pipes or damage to pumps.

screenings   (1) Material removed from liquids by screens. (2) Broken rock, including the dust, of a size that will pass through a given screen depending on the character of the stone.

screenings grinder   A device for grinding, shredding, or macerating material removed from wastewater by screens.

screw-feed pump   A pump with either a horizontal or vertical cylindrical casing in which a runner with radial blades, like those of a ship's propeller, rotates. See also *vertical screw pump*.

screw pump   A type of rotary pump in which liquid is carried between screw threads and displaced axially. These pumps can be open or enclosed (i.e., the flights may be encased in a cylinder).

seal water   Water used to provide a seal to prevent air from entering a pump along the pump shaft, excessive leakage at the pump shaft where it enters the pump casing, and contamination of the fluid being pumped. Seal water may be treated effluent or potable water.

sedimentation   (1) The process of subsidence and deposition of suspended matter or other liquids by gravity. It is typically accomplished by reducing the velocity of the liquid below the point at which it can transport the suspended material. Also called settling. It may be enhanced by coagulation and flocculation. (2) Solid–liquid separation resulting from the application of an external force, typically settling in a clarifier under the force of gravity. It can be variously classed as discrete, flocculent, hindered, and zone sedimentation.

self-priming pump   A pump that includes equipment to automatically fill the volute with the liquid to be pumped.

semi open impeller   A type of impeller that incorporates a single shroud, typically at the back.

septic wastewater   Wastewater undergoing anaerobic decomposition.

septicity   A condition produced by growth of anaerobic organisms.

settleable solids   (1) That matter in wastewater that will not stay in suspension during a preselected settling period, such as 1 hour, but settles to the bottom. (2) In the Imhoff cone test, the volume of matter that settles to the bottom of the cone in 1 hour. (3) Suspended solids that can be removed by conventional sedimentation.

shaft   A bar that is used to transmit motion to a mechanical part.

shear pin   A metal rod that is carefully designed for a specific piece of equipment to break when the load reaches a certain point. The shearing of this pin will automatically shut down the equipment before damage occurs.

sheave   A pulley with a grooved or v-shaped rim to hold round or v-belts.

shim   A thin piece of wood or metal used to align two components, for example, between the two parts of a bearing on a round shaft. In this case, as the bearing wears, shims are removed, allowing the two parts to be brought closer together and securing a better fit on the shaft. Shims are also used to space stationary components.

shutoff valve   A valve installed in a pipeline, typically a main line, to shut off flow in a section to permit inspection or repair. Also called a stop valve or isolation valve.

single-action pump   A reciprocating pump in which the suction inlet admits water to only one side of the plunger or piston and the discharge is intermittent.

single loop controllers   A self-contained unit that is comprised of microprocessor-based sequencers, timers, and programmable logic controllers that are all dedicated to controlling a single loop or specific system. These units are very reliable and can be tied into communication networks.

single-stage digestion   Digestion limited to a single tank for the entire digestion period.

single-suction impeller   An impeller with one suction inlet.

siphon   A closed conduit that lies above the hydraulic grade line, resulting in a pressure less than atmospheric and requiring a vacuum within the conduit to start flow. A siphon uses atmospheric pressure to continue or increase flow through the conduit.

siphoning action   Flow that is created, sustained, or increased using a siphon. See also *siphon*.

sleeve bearing   Also known as journal bearing, a metal jacket fully or partially enclosing a rotating inner shaft.

slime   (1) Substance of a viscous organic nature, typically formed by microbiological growth, that covers other objects. (2) The coating of biomass that accumulates in trickling filters or sand filters and periodically sloughs away to be collected in clarifiers. See also *biofilm*.

slinger ring   A metal or rubber ring that is placed on a pump shaft between the packing and the bearing. It protects the bearing by throwing water out away from the shaft as it rotates.

slope (1) The inclination of gradient from the horizontal of a line or surface. The degree of inclination is typically expressed as a ratio such as 1:25, indicating unit rise in 25 units of horizontal distance; or in a decimal fraction (0.04); degrees (2° 18 min); or percent (4%). (2) Inclination of the invert of a conduit expressed as a decimal or as feet (meters) per stated length measured horizontally in feet. (3) In plumbing, the inclination of a conduit, typically expressed in inches per foot (meter) length of pipe.

sodium hydroxide (NaOH) A strong caustic chemical used in treatment processes to neutralize acidity, increase alkalinity, or raise the pH value. Also known as caustic soda, sodium hydrate, lye, and white caustic.

sodium hypochlorite (NaOCl) A water solution of sodium hydroxide and chlorine in which sodium hypochlorite is the essential ingredient.

solvent A substance, typically a liquid, capable of dissolving or dispersing one or more other substances.

split bearing cover A type of bearing cover plate that is made up of two halves that can be easily removed to access the bearing for maintenance or inspection.

split case pump A kind of pump with casing that is split parallel to the drive shaft.

split rigid coupling A type of inflexible coupling that is split in half parallel to the pump shaft. This coupling is very useful when the pump impeller needs to be positioned further away or closer to the motor.

squirrel cage motor A type of polyphase motor that is widely used in water and wastewater treatment applications and can be purchased in many configurations. Enclosures for these motors include open drip-proof, totally enclosed non-ventilated, and totally enclosed fan-cooled.

Standard Methods (1) An assembly of analytical techniques and descriptions commonly accepted in water and wastewater treatment (*Standard Methods for the Examination of Water and Wastewater*) published jointly by the American Public Health Association, the American Water Works Association, and the Water Environment Federation. (2) Validated methods published by professional organizations and agencies covering specific fields or procedures. These include, among others, the American Public Health Association, American Public Works Association, American Society of Civil Engineers, American Society of Mechanical Engineers, American Society for Testing and Materials, American Water Works Association, U.S. Bureau of Standards, U.S. Standards Institute (formerly American Standards Association), U.S. Public Health Service, Water Environment Federation, and U.S. Environmental Protection Agency.

standby generator A generator that is maintained as a replacement power source in case of breakdown or malfunction. Typically fueled by gasoline, diesel, natural gas, or methane (sludge digestion gas).

standby motor An identical motor that is maintained in case of breakdown or malfunction of a pump's prime driver.

static head Vertical distance between the free level of the source of supply and the point of free discharge or the level of the free surface.

static level   (1) The elevation of the water table or pressure surface that is not influenced by pumping or other forms of groundwater extraction. (2) The level of elevation to which the top of a column of water would rise, if afforded the opportunity to do so, from an artesian aquifer, basin, or conduit under pressure. Also called hydrostatic level.

static suction head   The vertical distance from the source of supply, when its level is above the pump, to the center line of the pump.

static suction lift   The vertical distance between the center of the suction end of a pump and the free surface of the liquid being pumped. Static lift does not include friction losses in the suction pipes. Static suction head includes lift and friction losses.

stator   (1) The stationary member of an electric motor or generator. (2) Also, the stationary sleeve of the progressing cavity pump in which the rotor moves. The stator is typically made of an elastomer material.

strain gauge   A device that converts a movement, such as a coil of wire as it is loaded, into a change in electrical properties. This results in a very sensitive measurement of the strain. Strain gauges are often used in pressure sensors and weight scales.

stuffing box   A cylindrical recess, typically between the impeller and the radial bearing of centrifugal pumps, that houses four to six rings of packing and a water seal ring. Its purpose is to prevent the flow of liquid along the shaft.

submersible motor   A motor specifically designed to be operated while completely submerged in a fluid. Most submersible motors have internal cooling systems, special seals on power and control cable entries, and an integrated leakage detection system.

submersible pump   A motor and pump combination designed to be placed entirely below the liquid surface. Submersible pumps are generally constructed as pumps that are vertically closed-coupled to a submersible motor. They are used for domestic and industrial treatment applications and for pumping groundwater from wells.

suction gauge   A pressure gauge placed on suction piping near the point where the pump connection is made. The gauge reading gives a good approximation of the suction head conditions under which the pump is operating.

suction head   (1) The head at the inlet to a pump (see below). (2) The head below atmospheric pressure in a piping system. A suction head exists when liquid is pumped from an open-top tank and the liquid level is above the centerline of the pump. More commonly known as a flooded suction.

suction lift   The vertical distance from the liquid surface in an open tank or reservoir to the center line of a pump drawing from the tank or reservoir and set higher than the liquid surface.

suction valve   A valve that is located on the suction piping of a pump.

sulfate-reducing bacteria   Bacteria capable of assimilating oxygen from sulfate compounds, reducing them to sulfides. See also *sulfur bacteria*.

sulfur bacteria   Bacteria capable of using dissolved sulfur compounds in their growth; bacteria deriving energy from sulfur or sulfur compounds.

sulfur cycle   A graphical presentation of the conservation of matter in nature showing the chemical transformation of sulfur through various stages of decomposition and assimilation. The various chemical forms of sulfur as it moves among living and non-living matter is used to illustrate general biological principles that are applicable to wastewater and sludge treatment.

sump   A tank or pit that receives drainage and stores it temporarily, and from which the discharge is pumped or ejected.

sump pump   A mechanism used for removing water or wastewater from a sump or wet well; it may be energized by air, water, steam, or electric motor. Ejectors and submerged centrifugal pumps, either float or manually controlled, are often used for this purpose.

synchronous motor   An electric motor having a speed strictly proportional to the frequency of the operating current.

Teflon® grease   High-performance lubricant containing polytetrafluoroethylene (PTFE).

Teflon® tape   Tape containing PTFE used to lubricate and seal pipe threads.

torque   A force that produces or tends to produce rotation or torsion; also, a measure of the effectiveness of such a force that consists of the product of the force and the perpendicular distance from the line of action of the force to the axis of rotation.

total dynamic discharge head   Total dynamic head plus the dynamic suction head or minus the dynamic suction lift.

total dynamic head (TDH)   The difference between the elevation corresponding to the pressure at the discharge flange of a pump and the elevation corresponding to the vacuum or pressure at the suction flange of the pump, corrected to the same datum plane, plus the velocity head at the discharge flange of the pump minus the velocity head at the suction flange of the pump.

total head   (1) The sum of the pressure, velocity, and position heads above a datum. The height of the energy line above a datum. (2) The difference in elevation between the surface of the water at the source of supply and the elevation of the water at the outlet, plus velocity head and lost head. (3) The high distance of the energy line above the datum; energy head. (4) In open channel flow, the depth plus the velocity head.

total pumping head   The measure of the energy increase imparted to each pound of liquid as it is pumped, and, therefore, the algebraic difference between the total discharge head and the total suction head.

total static head   The vertical distance in height between the point of entry to the system and the highest point of discharge.

totally enclosed fan cooled (TEFC) motor   A motor constructed with a small fan on the rear shaft and typically covered by a housing. The fan draws air over the motor fins, removing excess heat and cooling the motor. Because the motor is totally enclosed, the motor is dust tight, and has a moderate water seal.

trap   (1) A device used to prevent a material flowing or carried through a conduit from reversing its direction of flow or movement, or from passing a given point. (2) A device to prevent the escape of air from sewers through a plumbing fixture or catch basin.

trash   Debris that may be removed from reservoirs, combined sewers, and storm sewers by coarse racks.

trash rack   A grid or screen placed across a waterway to catch floating debris.

turbine pump   A centrifugal pump in which fixed guide vanes partially convert the velocity energy of the water to pressure head as the water leaves the impeller.

turbulence   (1) The fluid property that is characterized by irregular variation in the speed and direction of movement of individual particles or elements of the flow. (2) A state of flow of water in which the water is agitated by cross currents and eddies, as opposed to laminar, streamline, or viscous flow. See also *turbulent flow*.

turbulent flow   (1) The flow of a liquid past an object such that the velocity at any fixed point in the fluid varies irregularly. (2) A type of fluid flow in which there is an unsteady motion of the particles and the motion at a fixed point varies in no definite manner. Also called eddy flow or sinuous flow.

ultrasonic sensor   A device that transmits and then receives ultrasonic waves that bounce off a liquid surface to measure the liquid level. The elapsed time required for the signal to leave and return to the sensor is a function of the distance to the liquid surface.

underdrain   A drain that carries away groundwater or the drainage from prepared beds to which water or wastewater has been applied.

universal joint   A joint or coupling that permits limited motion in any direction; used to transmit rotary motion between shafts that are not colinear.

universal motor   An electric motor that can be used on either an alternating or a direct current supply.

user   The party who is billed, typically for sewer service from a single connection; has no reference to the number of persons served. Also called a customer.

user charge   Charge made to recipients of wastewater services.

vacuum   A space condition that is characterized by pressure lower than atmospheric. This condition is created when air is pumped out of an unvented space.

vacuum breaker   A device for relieving a vacuum or partial vacuum formed in a pipeline, thereby preventing backsiphoning.

vacuum gauge   A gauge that indicates the pressure of the air or gas in a partial vacuum.

valve   (1) A device installed in a pipeline to control the magnitude and direction of the flow. It consists essentially of a shell and a disk or plug fitted to the shell. (2) In a pump, a passage through which flow is controlled by a mechanism.

valve seat   The surface on which the moving component of a valve rests while in a closed position.

valve throttling   To control or modulate flow through a system by manually or automatically opening or closing a valve to various degrees. In a pumped system, changing the discharge valve to various positions between full open and full closed regulates the amount of discharge head. Gate valves and plug valves are often used to throttle the discharge of centrifugal pumps.

vapor   (1) The gaseous form of any substance. (2) A visible condensation such as fog, mist, or steam that is suspended in air.

variable-frequency drive (VFD)   An electronic controller that adjusts the speed of an electric motor by modulating the power being delivered. VFDs provide continuous control, matching motor speed to the specific demands of the work being performed.

v-belt drive   A type of pump coupling that consists of a belt of triangular shape running on sheaves or pulleys with similarly shaped grooves. The movement of the belt in the groove creates a wedging action that improves the traction of the system and, therefore, gives it more pulling power than a round or flat belt system

v-belt drive pulley   It is a disk-shaped object that rotates around a fixed center. The outside edges or the disk are grooved to hold the belts. See also *sheave* as it is also known.

velocity head   (1) The vertical distance or height through which a body would have to fall freely, under the force of gravity, to acquire the velocity it possesses. It is equal to the square of the velocity divided by twice the acceleration of gravity. (2) The theoretical vertical height through which a liquid body may be raised by its kinetic energy. It is equal to the share of the velocity divided by twice the acceleration caused by gravity.

velocity meter   A veined water meter that operates on the principle that the vanes of the wheel move at approximately the same velocity as the flowing water.

Venturi meter   A differential meter for measuring the flow of water or other fluid through closed conduits or pipes. It consists of a Venturi tube and one of several proprietary forms of flow-registering devices. The difference in velocity heads between the entrance and the contracted throat is an indication of the rate of flow.

vertical pump   (1) A reciprocating pump in which the piston or plunger moves in a vertical direction. (2) A centrifugal pump in which the pump shaft is in a vertical position.

vertical screw pump   A pump, similar in shape, characteristics, and use to a horizontal screw pump, but which has the axis of its runner in a vertical position.

vibration sensor   An online sensor to monitor trends in equipment vibration to detect problems before failure.

viscosity   Measurement of a fluid's resistance to flow that reflects the molecular attractions within the fluid.

void   A pore or open space in rock or granular material not occupied by solids.

volatile   Capable of being evaporated at relatively low temperatures.

volute   One of the two main components of a centrifugal pump (the other is the impeller). The volute forces the liquid to discharge from the pump. This is accomplished by offsetting the impeller in the volute and by maintaining a close clearance between the impeller and the volute at the cut-water.

volute pump   A centrifugal pump with a casing made in the form of a spiral or volute as an aid to the partial conversion of the velocity energy into pressure head as the water leaves the impellers.

walking beam   A beam or lever pivoted at or near the center for transmitting motion by rods at each end.

wastewater   The spent or used water of a community or industry containing dissolved and suspended matter.

water column   (1) The water above the valve in a set of pumps. (2) A measure of head or pressure in a closed pipe or conduit.

water hammer   The phenomenon of oscillations in the pressure of water about its normal pressure in a closed conduit, flowing full, which results from a too-rapid acceleration or retardation of flow. Momentary pressures greatly in excess of the normal static pressure may be produced in a closed conduit by this phenomenon.

wear ring   A replaceable metal incorporated to pumps at the point where the clearance between the pump casing and the impeller are the closest. The wear ring can be placed on the casing, on the impeller, or at both locations. When these rings reach their wear limits, they can be replaced. This restores the design clearance on the parts and keeps the pump operating as efficiently as possible.

weir   A device that has a crest and some side containment of known geometric shape, such as a V, trapezoid, or rectangle, and is used to measure flow of liquid. The liquid surface is exposed to the atmosphere. Flow is related to the upstream height of water above the crest, position of crest with respect to downstream water surface, and geometry of the weir opening.

wet well   A compartment in which a liquid is collected, and to which the suction pipe of a pump is connected.

wire-to-water efficiency   The ratio of mechanical output of a pump to the electrical input at the meter.

worm gear reducer   A gear consisting of a worm (or short rotating screw) and a worm (or toothed) wheel.